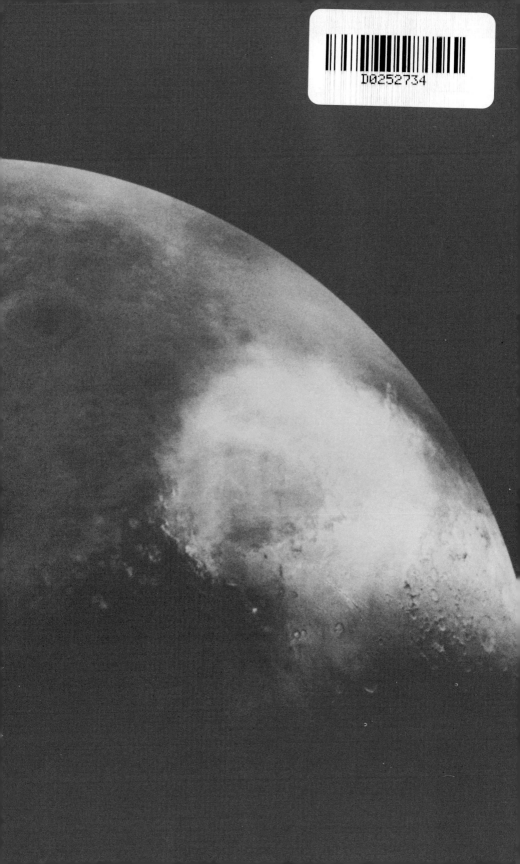

The
Explorers
of Mars Hill

The Explorers of Mars Hill

A Centennial History of Lowell Observatory

1894 ✣ 1994

William Lowell Putnam

and others

Published for

LOWELL OBSERVATORY

by

PHOENIX PUBLISHING

West Kennebunk, Maine

Library of Congress CIP Data

Putnam, William Lowell.
The Explorers of Mars Hill: a centennial history of
Lowell Observatory / by William Lowell Putnam and con-
tributors.
    p. cm.
Includes bibliographical references and index.
ISBN 0-914659-69-3
1. Lowell Observatory—History.   I. Title.
QB82.U62F537   1994
552'.19791'33—dc20   93-49057
                                                           CIP

Printed in the United States of America

verything around this earth we see is subject to one inevitable cycle of birth, growth and decay. Nothing begins but comes at last to end . . . though our own lives are too busy to even mark the slow nearing to that eventual goal . . . Today, what we already know is helping to comprehension of another world; in a not distant future we shall be repaid with interest, and what that other world shall have taught us will redound to a better knowledge of our own, and of that cosmos of which the two form part.

From Percival Lowell's
Mars and Its Canals
1906

.

Contents

Introduction

THIS IS the centennial story of the largest and best known privately managed astronomical observatory in the world, an institution founded and endowed by Percival Lowell in 1894. This is not, however, his biography, though we can recommend a forthcoming volume by David Strauss as by far the most thoroughly researched personal life of our founder. Herewith, we have assembled narratives of the more significant activities and scientific discoveries compiled over the years as a result of Lowell's interest and investment.

A bibliography of much material relevant to this history is given as a separate appendix (we have not attempted to list all those articles published in various scientific journals), thus a goodly part of this volume covers topics and material already in print about the work done at Dr. Lowell's institution in years past. Our part has been to focus on the personalities and to induce those more informed than we to provide meat for the layman's accounts that follow - a story of the notable astronomic discoveries which have continued to flow from this unique, privately owned and rigidly independent research institution.

We are obviously indebted to the great number of people who have worked on that "low hill west of Flagstaff," which was soon named "Mars," after the topic of Lowell's most popularized initial astronomic interest. The principal author is also deeply grateful for the help given with certain chapters that have been supplied almost entirely by others and for which specific credit is given in this narrative.

The value of this entire work would be greatly diminished without the sustained encouragement, historic perspective and strong supportive labor of Lowell Observatory's resident guru, Dr. Henry Lee Giclas, almost all of whose long life has been spent in association with it and from whose reminiscences, both oral and written, I am indebted for one whole chapter and numerous massive plagiarisms. The pictures used throughout this volume have been culled from among thousands in the

Lowell archives by the Observatory's all-purpose assistant, Mary Lou Evans, who has also provided kindly stimulus and generous assistance at too many turns for me to recall. These pictures have been made presentable by the diligence and competence of Paul Roques. To this has been added the ongoing pleasure of working with Lowell Observatory's outstanding director, Dr. Robert Lowell Millis, whose criticism and very substantial contributions have added much accuracy.

Certain other chapters have been contributed in toto by interested persons, friends and advisors to Lowell Observatory, whose more specific knowledge of events makes them authoritative. These include George Wesley Lockwood and Lawrence Harvey Wasserman, veteran staff astronomers at Lowell Observatory; Gibson Reaves, associate meritus of Lowell and professor of astronomy at the University of Southern California; William Edward Brunk, a longtime staff member of NASA, and Robert William Smith, Smithsonian historian of astronomy. With their cooperation, I have amended their submissions in order to achieve more conformance of style throughout this book.

Several generous "outsiders" have also helped, including Dr. Claudio Abächerli, of Visp, Switzerland, Arve Moen, Esq. of Trondheim, Norway, Dr. William Merz Sinton, astronomer of Flagstaff, David Strauss, historian of Kalamazoo College, Kathryn Flynn Broman, Anne Lockwood Seamans and David Chalmers Henley, valued members of Lowell Observatory's Advisory Board. I am further indebted to the Boston Athenaeum and the Boston Public Library for much assistance in my research.

Our founder had plenty of what Ferris Greenslet described as the Lowell characteristic of "animal spirits" and we have included a final chapter on apochrypha in which we ventilate a few exhibitions of this trait. Percival Lowell also made mistakes and, as is the fate of anyone who attracts public attention as much as he did, his mistakes got wide attention too.

This narrative is carried in considerable detail to within some dozen years of the centenary observation of Lowell Observatory—no further. The author has spent too long in the business of providing news and information to dare risk venturing onto the thin ice of recent events where perception of essential facts may well be biased by the cares of living egos. We will let more recent facts speak for themselves and leave any interpretations and insightful afterthoughts for the undoubtedly calmer view of our heirs.

As the current sole trustee under his will, I must confess to a high measure of pride in the necessarily close association with great-uncle Percy. But of far greater significance to any student of human nature is the self-evident fact that of those he assembled to work on Mars Hill, and the many who have followed in later years, the overwhelming majority have stayed close to, or in the employ of, his observatory throughout their entire scientific careers. We must be doing something right.

William Lowell Putnam

Mars Hill, Flagstaff, Arizona
April 1994

An Overview

THE HISTORY of the Lowell Observatory is a fascinating one. However, it tells of more than the exploits of a small band of astronomers and their telescopes. It is also a window onto a remarkable century in the history of astronomy, in particular, the swiftly changing nature of the astronomical enterprise in the United States. Nevertheless, observatories are not islands separate from the general culture, but rather form part of it. Broad social shifts helped to shape the activities of astronomers atop Mars Hill, just as the findings of those astronomers reached out not only to their colleagues, but to the wider society too. In this short overview, we will take a step back to try and place the history of Lowell Observatory into the broad picture, a picture that should be borne in mind as we delve into more detailed accounts in later chapters.

For the first half of the nineteenth century, American science counted for little on the world stage. But in mid-century, this started to change. In the late 1840s, for example, two American astronomers engaged in a fierce debate with European astronomers over the allocation of credit for the discovery of the planet Neptune. While both sides were somewhat bloodied in the fight, it certainly demonstrated the talents and abilities of American scientists in an extremely demanding branch of what was widely regarded as *the* science—astronomy.

But like American science in general, American astronomy was in the process of establishing the foundations of a mature community at the time of the Neptune episode. The most striking feature of this development was the proliferation of observatories. In the fifteen years preceding the discovery of Neptune, for example, observatories were established at the University of North Carolina, Yale, Williams College, Western Reserve College, West Point, Philadelphia Central High School, Harvard, Cincinnati, Georgetown, and Washington, DC (U.S. Naval Observatory).

While these new institutions depended largely on European instrumentation through the 1840s, the observatory movement in America did help spawn domestic telescope manufacturing, in addition to expanding employment and research opportunities for American astronomers. Similarly, two astronomical journals, the short-lived *Sidereal Messenger* as well as an *American Nautical Almanac*, were launched in the late 1840s.

Nevertheless, writing in 1874 the leading American astronomer, Simon Newcomb, was still acutely aware that Europeans were far ahead. Newcomb examined the state of the exact sciences—which he took to be astronomy, physics and its ally mathematics—in the United States. As Albert E. Moyer has put it,

Contrasting the quantity of research published by Americans and Europeans in these fields, he surmised that the Americans lagged far behind their overseas colleagues in level of achievement by a generation. He cautioned that this lag was not because of inadequate facilities or unqualified personnel, especially in the eastern states, researchers had access to good libraries as well as expensive instruments and apparatus. And the country possessed its share of first-rate practitioners in physics and astronomy. In actuality, Newcomb explained, the lag correlated with deficiencies in research journals, universities, and professional societies.[1]

For Newcomb the fact that basic research was held in low esteem in the United States led to the poor infrastructure for research. This had to be changed:

The remedy is to educate the intelligent public into an appreciation of the importance of scientific investigation, and of the necessity of bestowing upon those who are successfully engaged in it something in the way of consideration which may partially compensate them for devoting their energies to tasks which, from their very nature, can bring them no pecuniary compensation.[2]

Americans were not quite so indifferent to basic science as Newcomb believed. Such appeals by astronomers and their allies bore fruit. In fact, the United States would shortly become a leading power in observational astronomy and astrophysics, and by the end of the first decade of the twentieth century, the dominant power. This rise mirrored

[1] Albert E. Moyer; *A Scientist's Voice in American Culture*; U. of California Press, Berkeley, 1992; p. 83.
[2] Moyer; Op Cit, p. 84.

the country's climb to economic power. Nor, as Howard Miller has argued, were these two developments independent of each other since the prosperity of American manufacturing industries, along with burgeoning banking and transportation concerns, thrust fantastic fortunes into the hands of a small number of entrepreneurs and landowners. Some of these people donated part of their wealth to the arts and sciences and the money of this elite often went to endow astronomical and astrophysical research; the observatories of Lick, Yerkes and Mount Wilson are all monuments to scientists who managed to wrestle successfully with the wealthy and the generous, and much of the research conducted at the Harvard College Observatory, for example, was funded by private philanthropy. Hence ". . . astronomy and astrophysics, two of the least immediately utilitarian of the sciences, became the most richly endowed of all the sciences in the United States."[3]

But those who gave their support to astronomical and astrophysical enterprises did not engage in such researches themselves. A striking exception was Percival Lowell.

The wealthy Boston patrician and member of the United States untitled aristocracy, in this, as in many other things, did not play by the usual rules. The observatory he founded at Flagstaff in 1894 was his own. He answered to no one. Also, at a time when amateur astronomers were being increasingly outgunned by the more powerful telescopes and sophisticated pieces of auxiliary apparatus of the professionals, Lowell and his small staff—never more than ten during his lifetime, and usually smaller—were able for the most part to exploit state-of-the-art tools, most notably the very fine 24-inch Clark refracting telescope, first erected at Flagstaff in 1896. And when Lowell decided in 1900 that he needed a spectrograph for this telescope he demanded of its maker that it be the best of its sort in the world. In addition, at a time when astronomers were turning increasingly to the study of objects far beyond our Sun and its planets, Lowell focussed his own and his staff's attentions on the constituents and evolution of the solar system.

Through his use of the Clark refractor and the claims that sprang from these observations, Lowell thrust himself to the forefront on a range of issues, most emphatically the debate on the existence of life on Mars. Lowell did not start this debate but he became its most public participant. For Lowell, the markings he glimpsed criss-crossing the planet's surface were canals. These, moreover, were interpreted by Lowell as the product

[3]H. Miller, *Science and Private Agencies*, pp. 191-221, quoted on p. 203 in M. Hall and D. van Tassell (eds) *Science and Society in the United States*; Dorsey Press, Illinois, 1966.

of an advanced civilization that sought to bring water from the planet's polar regions to irrigate its deserts. His espousal of such views drew heated condemnations from some professional astronomers. What got under the skins of these professionals was not just that Lowell advanced controversial views, but that he argued for them with such energy and tenacity in arenas far outside those in which scientists usually thrashed out their disagreements. Lowell, who felt no ties of deference to the power brokers of American astronomy, went over the heads of the professionals and appealed directly to the general public in a barrage of lectures and vividly written articles and books. What is more, the subject engrossed the public and Lowell was a skilled polemicist. While he enjoyed cordial relations with many professionals, to his opponents, dealing with him seemed more like a political campaign than a scientific exchange of views.

In the eyes of some of the leaders of the professionals in the United States, in particular the powerful George Ellery Hale of Yerkes then Mount Wilson and W. W. Campbell of the Lick Observatory, Lowell put himself beyond the pale. In their view Lowell threatened to scupper their own plans to advance astrophysics. So while Lowell and his observatory were soon very prominently in public view, their reputation in the eyes of some professionals was low.

When in 1913 one of Lowell's staff members, Vesto Melvin Slipher, reported to Campbell on some seemingly astonishing results he had found for the speed of an astronomical object, Campbell was highly skeptical. But Slipher, Campbell and others would soon agree, was right; as Campbell wrote, Slipher's results ". . . compose one of the greatest surprises which astronomers have encountered in recent times."[4] In fact, Slipher had secured, and for the next decade or so would continue to secure, results on the speeds of galaxies that would later be interpreted as key evidence of an expanding universe (as will be seen in detail in Chapter 4). By any standards, Slipher's results were among the most important in twentieth century astronomy and had demanded from him an extremely high level of skill indeed. As nothing had before, Slipher's results announced that researches conducted at Lowell Observatory had to be taken very seriously.

But despite such spectacular finds as Slipher's results on the speeds of galaxies, Lowell's proselytizing on behalf of his views on Martian canals and intelligent Martians left a legacy for the Observatory that reached far beyond his death in 1916. In fact, his loyal and longtime band

---

[4]W. W. Campbell to V. M. Slipher, 2 November, 1914; Lowell Observatory Archives.

of assistants—V. M. Slipher, E. C. Slipher and C. O. Lampland, who would between then rack up nearly one hundred and fifty years of service to the Observatory—while holding him in the highest respect and being convinced of the essential correctness of his views on extraterrestrial life, strenuously sought to avoid taking any strong public positions on theoretical issues in astronomy for fear of criticism. The literary fisticuffs Lowell indulged in were anathema to the Sliphers and Lampland. Instead, they labored to secure observational results that they judged to be beyond reproach. The interpretation of these results, and the disputes that might arise, could be left very largely to others. Thus the Observatory's programs, such as for example, E. C. Slipher's series of photographs of Mars or Lampland's measurements of infrared emissions from planets, emphasized long-term projects and efforts to improve observational techniques, not theoretical innovations.

Lowell left another legacy. After his death the staff largely continued lines of research that Lowell had initiated or encouraged. One such investigation was the search for Planet X, the existence of which Lowell had become convinced of through extensive calculations (his Harvard teachers regarded him as a brilliant mathematician). As we will see in Chapters 9 and 10, in 1930 these researches were crowned triumphantly with the discovery of Pluto.

First class research did continue at the Observatory after Lowell's death—the discovery of Pluto being the most obvious example—but in comparison with leading observatories, the most significant being Mount Wilson and Lick, it went into relative decline. There were a number of reasons for this. Some of them were local, while others had more to say about attitudes towards the funding of scientific research in the United States. In the late 1910s at the same time that the world's most powerful telescope—a 100-inch reflector—was being put into operation atop Mount Wilson in California, the Lowell Observatory was being dragged ever deeper into a long, grinding and financially devastating battle with Percival's widow over the fate of his estate. When, more than a decade after Lowell's death, the fight was over, the upshot was that the Observatory's income was much lower than Lowell had intended. In fact, from 1916 (when V. M. Slipher became first acting director) until the late 1940s, the observatory was often strapped for cash, and in resources, people and equipment was increasingly left behind by larger and better funded institutions. That the Observatory survived at all was an achievement and owed much to its trustee from 1927 to 1967, Roger Lowell Putnam, who at times shored up the institution with his own money.

That the Observatory was in a perilous position by the late 1940s was undeniable. All but one of the younger staff had been released, Lowell staff members very rarely appeared at professional meetings, and almost no scientific papers were being published. However, at the same time, federal funds had started to flow to science far more readily than ever before. Scientists had demonstrated how crucial they were for national defense in World War II, and this, along with mounting Cold War pressures (and later the Korean War and then in the late 1950s the Sputnik crisis), opened federal purse strings to an unprecedented extent. And it was from federal sources that the revival of the Observatory sprang. What had been impossible before the war was brought within reach. With the arrival in 1958 of John Hall as director (he served as such for nineteen years), the institution had secured not only a front-rank astronomer, but also someone whose skills in attracting federal dollars and first rate colleagues were to be the key to the Observatory's continuing revival and transformation.

Most significantly for Lowell Observatory, following the establishment of NASA (the National Aeronautics and Space Administration) in 1958, funds were available for the study of the solar system as never before; the new space agency in fact would play a key role in the rapid growth of planetary science. Before the war, the great majority of astronomers, including the Lowell staff, worked hard for grants of a few hundred dollars; by the late 1950s, astronomers were quite used to thinking in terms of tens and hundreds of thousands, and for some enterprises, millions of dollars. With the accumulated expertise, knowledge and interests of its growing staff in the solar system, this observatory was well positioned to take advantage of this new source of patronage. The changed relationship between the Observatory and the federal government was symbolized most clearly in the new Planetary Research Center, opened in 1963 and made possible by NASA. But the interests of the growing Lowell staff were not restricted solely to the solar system. With the arrival in 1963 of William Baum, for example, the Observatory had recruited not only an expert on solar system astronomy, but also someone who was a top rank researcher in studies of galaxies. There were also different approaches to research, in particular, more cooperative enterprises. During the 1950s and 1960s, the Observatory was in many ways remade with the arrival of new staff, new equipment, new buildings, new forms of support and new ways of doing business.

In its renewal with substantial infusions of federal funds, the history of Lowell Observatory mirrors quite closely the history of much astronomy in the United States during the post World War II era. But by the late 1980s, the Observatory's leaders had become increasingly concerned that their institution was perhaps too dependent on federal monies. With roughly two-thirds of its income coming from this source, and with increasing demands on federal dollars for science, there began a concerted attempt to look to the Observatory's future by seeking out new funding opportunities. As Lowell Observatory enters its second hundred years, the need to continue to remake it in the light of new scientific and technological opportunities, as well as wider shifts in the funding of science in the United States and the changing place of science in American life, presents it with challenges that while new in their particulars have also been present, as noted in the rest of this book, in various ways during its first century.

Robert Smith

Washington, D. C.
1 June, 1993

Produced by the art students at East Flagstaff Junior High School in 1980 under the direction of artist and teacher Steve Nelson, this 5'x4' acrylic painting now hangs in the rotunda of the Slipher building at Lowell Observatory.

The
Explorers
of Mars Hill

I

Lowell Observatory
Gains Stature

The Clark telescope with the new Brashear spectrograph attached, 1903

1

The New World
1634 – 1894

P ERCIVAL LOWLE (1571-1664)
was born in Somerset, England,
a member of the local business
community with shipping interests in the nearby Severn estuary port of
Bristol. He also served as an assessor for the local Earl of Berkeley, to
whose heir, ten generations later, one of his descendants (Mary Emlen
Lowell (1884-1975)) would become married. Though moderately well-to-
do, Percival decided to emigrate to the New World with his grown sons,
as reports were now common that it was a land of greater opportunity
than old England.

An eighth generation descendant of Percival, though not in the line we
will trace, was the Reverend Delmar Rial Lowell (1844-1912), of Rutland,
Vermont, who lost his right arm in the Civil War and became chaplain-in-
chief of the Grand Army of the Republic. After twenty-five years of
research, he compiled an eight-hundred-page genealogy of the Lowell
family[1] wherein he notes that the name first appears in the Battle Abbey
records of early Normans who came to England with the Conqueror in
1066. The name reappears in 1220, attached to one William Lowle of
Yardley in Worcestershire and can be traced downward from him
through ten further generations—to the first Percival.

When close to his seventieth year, Percival sailed on the *Jonathan* in
1639, and settled in the town of Newbury, some thirty miles north of
Boston, where he took up land and his sons established themselves as

[1] *The Historical Genealogy of the Lowells of America from 1639 - 1891.*

3

Abbott Lawrence and Percival Lowell in childhood, 1859

farmers. Nine generations later, via six Johns, one Ebenezer and finally Augustus, yet another Percival Lowell was christened in Boston in the spring of 1855. (See Appendices)

In 1854 Augustus Lowell married Katharine Bigelow Lawrence (1832-1895) daughter of Abbott Lawrence, his father's neighbor and business associate, and at that moment United States minister to the Court of St. James. The wedding created a further bond among those making up the "establishment" of Yankee fortunes. Lawrence (1792-1855) was the most prominent member of another family whose name is on the Massachusetts landscape in association with American industrial development.

Over the next twenty years the new couple's adult children were Percival (1855-1916), Abbott Lawrence (1856-1943), Katharine (1858-1925), Elizabeth (1862-1935) and Amy (1874-1925). They each followed distinct lines of endeavor which led most of them to prominence in different fields.

A. Lawrence Lowell became a member of the Massachusetts bar, undertook the study of government and became the leading authority on the still unwritten constitution of Great Britain. As a lecturer in Government at Harvard, he achieved such distinction that he was elected President of the University in 1909 upon the retirement of Charles Eliot. Soon after assuming this office, Lawrence noted that there appeared to be two evolving lines of education, the generalist and the specialist—the former appearing to be "learning less and less about more and more, until finally they know nothing about everything." Whereas the specialists were tending to learn "more and more about less and less until finally they knew everything about nothing." To avoid the possibility of an imbalance in higher education, he established the concept whereby certain minimum scholastic achievements were required in several topics for graduation, as well as a field of concentration. He was also responsible for the development of university-operated dormitory facilities in place of the cliquishness that many perceived in the fraternity system. During his reign Harvard doubled its student enrollment, tripled its faculty size, established three new graduate schools and quintupled its endowment.

Katharine married Alfred Roosevelt in 1882 but was widowed in 1891, leaving her with three young children. Ten years later she married Thomas James Bowlker (1858-1917) and resettled to England, where her elder daughter, Elfrida, married Sir Orme Clarke. Her younger daughter, Katharine, married Joseph Stanley Reeve and lived for many years in Pennsylvania.

Elizabeth, the second daughter, was married in 1888 to her third cousin, William Lowell Putnam, eldest son of George Putnam (1834-1912), a law partner of William Russell. Soon after graduation from Harvard in 1882, William (1861-1924) joined his father's Boston firm which, after several more partnership changes over the years, is still to be found in legal references as "Putnam, Putnam & Bell," though at this writing not one of those surnames can be found on the firm's masthead.[2] A director of many companies, chairman of three and counsel for the burgeoning Bell System, William also showed the by now legendary Lowell talent for mathematics and chaired for many years the Visiting Committee for Harvard's Mathematics Department.

For her part, Elizabeth took a notable part in the affairs of the day. Spurred by the death in 1900 of her three-year-old daughter, Harriet, she led her state and the nation in a campaign for cleanliness standards in milk production and distribution. She organized a social program for the orderly "Americanization" of immigrants and, with the vigorous assistance of her daughter-in-law, Caroline, took a prominent part in achieving repeal of the Eighteenth Amendment and the ineffective and harmful Volstead Act. In addition, many people said that her privately printed volumes of poetry were superior to the more famous works of her "little" sister.

Amy, known within the family as "the postscript," was named for her great aunt Rebecca, daughter of John Lowell "the Rebel," who had died only a few weeks before her birth. Eschewing the married state, Amy opted for a career of writing, poetry and public speaking. She became the family "character," competing for attention distinct from her more famous brothers by wearing men's clothing, smoking cigars, and after replacing her carriage with one of the horseless variety, even driving her own car around the countryside. Physically overweight but widely admired for her poetic works, Amy lived out her entire life in the home of her birth—Sevenels.

This narrative follows the most famous activity of the eldest sibling, the second Percival. This work was aided by his brother-in-law, William Lowell Putnam, II,[3] and his brother, Lawrence; persevered in through painful adversity by his third cousin, Guy (1870-1927); doted on for forty

[2] The Law firm of "Putnam, Bell & Russell" still operates in Chatham, Massachusetts, with Charles Russell as its senior partner.
[3] There have been four adults bearing this name, but not in genealogical sequence. The first died at age 21, of wounds suffered at the Battle of Balls Bluff in 1861 and the name was given by his cousin to her first-born son a month later. The other three have all had a relationship with Lowell Observatory. In this text, where necessary, they are distinguished by numerals.

years by his nephew, Roger Lowell Putnam (1893-1972), thereafter by his sons, Michael and William, and executed by the distinguished scientists assembled to work under Percival's inspired—even if on one topic gloriously flawed—leadership for a century of astronomical discovery on what soon became very appropriately known as Mars Hill.

Sevenels was the name that Augustus applied to the home he and Katharine acquired in 1867 on the corner of Heath and Warren streets in the then fashionable and still rural Boston suburb of Brookline; there were seven persons living under that roof whose name began with the letter "L." The couple's first residence after marriage had been near her father's home on Beacon Hill, at 131 Tremont Street, between Winter Street and Temple Place. This spot was later occupied by the Shepard Department store, whose proprietors were among the nation's pioneer broadcasters. Even later, this real estate, adjacent to the Episcopal Cathedral Church of Saint Paul, was occupied by an eight-story office building housing much of the Massachusetts Attorney General's office.

Because of Katharine's delicate health following the birth of five children (Elizabeth's twin brother, Roger, had died at sixteen months), as was common among the Lowells, Augustus took his family for an extended European visit in 1864. Though the parents were based primarily in Paris, the two boys were sent to a nearby school run by one Kornemann from which they were allowed home only on Sundays. In this school Percival learned the local tongue by the technique later called immersion, and developed an aptitude for languages that served him well throughout the rest of his life.

Traveling around "the continent" during their two-year absence from Boston, the family became witnesses to an episode in the "Seven Weeks War," one of the countless brief conflicts by means of which the numerous petty German states became united under the ruling Hohenzollern family of Prussia. Percival, however, showed a distaste for travel (from which he obviously recovered later in life) and, in the spring of 1866, was placed in the famous school run by Frederick Edouard Sillig (1804-1871) at Bellerive La Tour, near Montreux in Switzerland, while the rest of the family went on a tour of Italy.[4]

By late that summer everyone was back at Sevenels and Percival was

[4]Young Percy was in good company. Sillig's class of 1866 included Victor Wilson of New York; Chase Harrington, who became Minister Plenipotentiary to Switzerland; Richard and Charles Freedman, later investment counselors of Boston; Alfred Tysoe of Manchester, England, who became an Anglican priest; and Arthur Donner of Frankfort am Main.

Percival Lowell stands next to his Japanese companion, "Tejiro" (see photograph on page 88), with three Western friends at his right.

sent to Mr. Noble's (Classical) School, preparatory of course for Harvard College, and run by a man, George Washington Copp Noble, whose name is distinguished in American secondary education.[5] But the teenager had also acquired a $2\frac{1}{4}$-inch refracting telescope, by means of which he was able to distinguish the changing coloration of the polar regions of the planet Mars from the rooftop of Sevenels.

Entering college with the class of 1876, Percival followed tradition and excelled in mathematics, though he also won a Bowdoin[6] prize for his essay on "The Rank of England as a European Power from the Death of Elizabeth to the Death of Anne." As a Phi Beta Kappa graduate, he met the family's self-imposed traditions of scholarship, intellectual discipline and attainment. His graduating thesis—a portent of things to come—was an essay on "Nebular Hypothesis."

[5]Noble (1836-1912) was an archtypical Mr. Chips, whose successors are still at it as Noble & Greenough Academy in Dedham.
[6]James Bowdoin (1726-1790) was a trustee of Harvard College but gave his name to another, and was the first president of the American Academy of Arts and Sciences. He was also the governor of the Commonwealth whose painful duty it was to suppress the debtors' rebellion of 1786-7 led by former army captain, Daniel Shays.

Percival Lowell dining in his Tokyo garden with Ralph Curtis in 1883

Accompanied by his college classmate, Harcourt Amory, Jr., Percival then undertook an extended tour of Europe, during the course of which, after his solo journey down the Danube, the two visited Syria and Palestine, gaining a rare—for those times—insight into the mores of peoples that were seldom considered worthy of attention in contemporary European and American culture.

After his return home in the summer of 1877, Percival worked for six years in the office of his grandfather, John Amory Lowell, learning the myriad details of managing the family's business interests. Here he served for a time as treasurer[7] of the family's major cotton mills in the eighty- year-old city of Lowell.

In the spring of 1883, a year after the death of his eminent grandfather, Percival threw it all up and went to Tokyo with his telescope. There he rented a house and again showed his linguistic versatility by learning the Japanese language so effectively that, after only a few weeks of residence, he felt able to comment on its grammar in a letter to his mother. By mid-

[7]Under then-prevailing Massachusetts corporate law, the treasurer was elected by vote of the stockholders, not the directors, and thus was the most responsible of all officers.

9 / *The New World*

The first Korean trade mission to the United States after arrival in San Francisco in 1884. From left are Vice-ambassador Hong Yong Sik, Percival Lowell, Counsellor Shee Kwan Pan, and Ambassador Ming Yon Ik.

summer he was held in such esteem that the United States Legation asked him to serve as Foreign Secretary and Counsellor to a three-man Korean delegation about to visit the United States. At that time, Korea was effectively dominated by Japanese concessions and its foreign affairs were largely out of local control.

In early September Percival found himself back in the United States, shepherding his small flock across the nation for six eventful weeks including a meeting with President Chester Arthur in New York. The delegation having achieved a number of their objectives, Percival escorted them home and was then invited to spend the western celebration of Christmas at the royal palace in Seoul as guest of the king.

Leaving Korea in early spring of 1884, Percival returned home by way of Shanghai, Hong Kong and Bombay (where he visited with his cousin Charles, then employed as the local agent of a French bank). Thence he continued to Venice, Paris and London, arriving in Boston before year end. At home again, he devoted himself to writing about his Far Eastern experiences, producing numerous articles for the *Atlantic Monthly* which were subsequently compiled into two books.[8]

Back in Japan towards the end of 1888, he stayed until the following spring and continued to write articles which he sent home for publication. Several of these articles were compiled into "NOTO—An Un-explored Corner of Japan" which appeared in 1891. Those connected with Lowell and his observatory who have visited the city of Animazu and the Noto area in more recent years have found that the memory of his visit, albeit very brief, is very much intact and still highly respected. Signs indicating landmarks of his visit can be found in several locations and a local bakery even offers "Percival Lowell cakes."

In 1890 he went again to Europe, visiting Spain for the first time; but the spring of 1891 found him back in Japan. This time he was accompanied by his friend and Harvard associate, George Russell Agassiz (1862-1951) (zoologist grandson of Louis Agassiz, the great Swiss-American naturalist) who was later a frequent visitor at Flagstaff. Touring the central region of Honshu, they visited a sacred site, the extinct, rounded volcano of Ontake, where Lowell came into close contact with the depth of religious fervor associated with Shintoism.

These latter two visits resulted in another book, on *Occult Japan*. But Noto was his most long-lasting impression. So deeply did Percival identify himself with that peninsula that a century after his visit there the mayor and other local dignitaries came to Mars Hill to pay their respects

[8]*Choson, The Land of Morning Calm* and *The Soul of the Far East.*

11 / *The New World*

at the grave of the man who had brought prominence to their region.

During these several years of travel, Percival's long-standing interest in astronomical questions came to the fore. In 1892 he had acquired a 6-inch telescope, much larger than the instrument of his teenage years, but still readily portable. With this he made a series of observations of Saturn and turned his attention again to Mars. Coincidentally, in that same year, Giovanni Virginio Schiaparelli (1835-1910) announced his retirement from astronomical observation because of failing eyesight. For twenty years Schiaparelli had been director of Milan's Brera Observatory[9] and was notable for his descriptions of lines on the Martian surface which he called "canali," the Italian word for "grooves." This word had been commonly but erroneously taken straight into English as "canals." In 1872 Schiaparelli had been awarded the Gold Medal of the Royal Astronomical Society for his studies of Mars.

Home again at the end of 1893, Percival Lowell, financially secure, well read and scientifically curious, began his move to take up the slack caused by the retirement of the distinguished Italian astronomer and delve into the study of Mars with its apparently strange markings. For this purpose, he would need an observing locale free from the bright lights of civilization, the haze of atmospheric humidity and the Eolian turbulence often found in rough terrain.

[9]Located near the present heart of Milan, this is no longer a good spot for observing. Brera (ancient Lombard for cultivated field) was started in 1760 as part of the Jesuit College which had been established in the Spanish Palazzo of Brera in 1671 under the patronage of the great Cardinal Carlo Borromeo. Borromeo, scion of a locally distinguished family, had the clout to make a notable move in science - 146 years after Pope Urban VIII forced Galileo to recant his theories on the Sun, stars and planets.

Giovanni Virginio Schiaparelli, circa 1870

2

Unto The Hill
1894

THE ASTRONOMY Department of Harvard University was effectively established with the legacy given to America's oldest institution of higher learning by Uriah Atherton Boyden (1804-1879) who left more than a quarter of a million dollars to get things moving. Boyden had been a surveyor, engineer and inventor who laid out numerous railroad lines in New England but made his mark financially by designing water-driven turbines for use by the numerous mills along the Merrimack River. Both the mile-long Amoskeag Mills of Manchester, New Hampshire, and the Appleton Mills of Lawrence, Massachusetts, owed their motive power to Boyden's turbines.

The principal personal beneficiaries of Boyden's legacy were the brothers Pickering—Edward Charles (1846-1919) and William Henry (1858-1938)—grandsons of George Washington's first Indian Commissioner and nephews of Charles Pickering (1805-1878), a naturalist famous in American scholarship and science. Both of the younger Pickerings were to leave their marks in scientific progress as a result of Harvard's new-found ability to further astronomic research.[1]

Though urban smog had yet to be diagnosed, beginning in 1886 Harvard astronomers were actively searching for observing sites away from the low altitude, high humidity, industrial smoke and encroaching lights of New England's cities. Even more relevant to this narrative, the younger

---

[1] For analysis of the relationships between the brothers Pickering and Percival Lowell, see the article by David Strauss in the winter issue, 1994, of the *Annals of Science*.

Pickering had devised a scale by which to judge and rate the precious and elusive quality of the critical astronomical condition known as "seeing." As in later scales of the ideal, it ranged downward from a perfect ten.

Stars appear to twinkle, not because of poetic need or an inherent fluctuation in the light emanating from these far distant sources, but because of the passage of that light through the turbulence of the Earth's atmosphere in the last few milliseconds of its long journey through space. To astronomers, this turbulence is, at best, a cause of great annoyance. Atmospheric fuzziness is what has driven astronomers away from the comforts of civilization and to remote mountain sites in areas of dry climate and more recently, to send their instruments into Space. Space has numerous additional advantages, among which is the possibility of viewing radiation in the ultra-violet portion of the spectrum, which is almost completely filtered out by the Earth's atmosphere.

In 1887, ten years after assuming the directorship of the Harvard Observatory, the elder Pickering brother had asked his associate in alpinism, Charles Ernest Fay (1846-1931),[2] a tenured professor of modern languages at Tufts College, to undertake a trip around North America with John Rayner Edmands (1849-1910), a fellow astronomer at Harvard.[3] They were asked to seek out and inspect possible observatory locations in the West. As a result of their report, W. H. Pickering spent the winter of 1888-89 on Mount Wilson in California, making the first observational use of a place later to become internationally noteworthy.

Searching also for a site whence the southern half of the sky would be more fully visible, in December 1890 E. C. Pickering sent his brother to South America in charge of a further search for observing locations. Going along on this trip as an assistant was Andrew Ellicott Douglass (1867-1962), a recent graduate of Trinity College where he had majored in astronomy and had then sought employment of exactly this nature. After checking the seeing at locations in Chile, Pickering and Douglass settled on a spot in Peru some eight thousand feet above sea level near Arequipa. Here they established Harvard's first "Boyden Station" and spent most of the next two years principally in the study of binary stars and nebulae. They also initiated a photographic program that would be continued for most of the next eight years and result in thousands of plates showing the spectra of eleven hundred Southern Hemisphere stars. Carrying much of

[2]Fay and the elder Pickering had been among the founders of the Appalachian Mountain Club in 1876 and Fay was on his way to becoming the godfather of North American alpinism.
[3]Edmands' name is recalled with reverence by New England hikers for the quality of the trails he laid out and built across New Hampshire's Presidential Range.

their unreduced data with them, Douglass and Pickering returned to the United States in mid-1892.

Mars is not the nearest of Planet Earth's fellow-travelers in orbit around the Sun, nor is it the most comparable to Earth in several other ways. Those honors belong to Venus, our "sister" planet of very similar size, which orbits next nearer the Sun (in Lowell's time its surface remained unknown, being hidden behind the planet's murky, sweltering atmospheric veil). But after Venus, Mars is the most Earthlike of the planets and in many ways the most easily observed. In its orbital circuit (synodic period) of 687 Earth days, Mars occasionally comes sufficiently close to Earth that these weeks of greater proximity, called oppositions, are events sought after by astronomers who wish a close-up view. (See page 223 for more on oppositions.) Such a condition was due to occur in the summer of 1894 and Percival Lowell, grasping the mantle of Martian study from the retiring Schiaparelli, determined to spend his time and talent in pursuit of information about this planet. He also put his money where his mouth was.

Early that year Lowell engaged W. H. Pickering and his now highly qualified assistant, Douglass, to help determine for him an optimum site for an astronomical observatory. Pickering suggested checking out the American Southwest, where generally dry air combined with high altitude to meet some of the elusive criteria associated with good observing conditions.

Armed with Pickering's scale of seeing quality and Lowell's now well-traveled, six-inch telescope, Douglass was sent to Arizona at the end of February 1894 via the newly opened Southern Pacific Railroad. His orders were to check out the observational possibilities at several locations in the Arizona Territory - Tombstone, Tucson, Tempe, Prescott and Flagstaff.

Douglass stepped down from the train at Benson, Arizona, early on March 8th and immediately took the stage to the already famous silver mining town of Tombstone. Here he contacted A. H. Emanuel, proprietor of the Aztec House and clerk of the local court, who showed him considerable hospitality. That night Douglass focussed the telescope on Jupiter and initially found the seeing to be at ten, but then a breeze came up and conditions deteriorated sadly.

According to the Tombstone *Prospector,*

Every man, woman and child were allowed a peep at Jupiter and Saturn. The latter with its rings was a most beautiful sight, as was the former with its four moons, auxiliaries which can not be seen without the aid of a powerful glass.

The next day, Mayor George Fitts, a prominent merchant, told Douglass the city would donate land for any observatory and build a road to it if necessary. The clerk offered free lodging for all visiting astronomers and everyone assured him that despite the desperado image of daily shootouts that was held elsewhere of this locale, the town was safe and the citizens law-abiding. Douglass passed this information on to Lowell and spent the next two nights testing further seeing in the Tombstone area. One of his observing spots was on the Grand Central Mine dump (the area's largest) and another on Reservoir Hill. At both locations, he found conditions to be good but not excellent, largely because of turbulence induced by the wind. He left for Tucson on the 11th and the *Prospector* reported:

The question of location will be settled inside of ten days. The location of the observatory means the erection of a large frame structure and the employment of a number of men. Connected with the institution will be three scientists. This is a small part of the benefit to be derived, however. Indirectly Tombstone will benefit greatly from its location here. People will travel from all parts of the world to inspect the heavens through the mammoth lens. Invalids will be drawn hither by the announcement that Tombstone has the clearest, dryest atmosphere with a less number of cloudy days than any point in the United States. The attention of medical men will be turned our way. It will be an advertisement worth thousands upon thousands of dollars yearly to Cochise county . . .

Without telling tales out of school it may be said that nothing has been left undone that could have been done to induce the gentleman at the back of the enterprise to erect the observatory at this point.

The next night, Douglass began a series of observations on high points of the Tucson area. Here the Tucson *Citizen* followed his activities with equal interest, noting under date of 14 March, 1894:

Professor Douglass made observations on the hill south of town just below the Julius Flynn stone quarry . . . The Professor had an hour or two of good "seeing." The site was not high enough to satisfy the professor. This afternoon he rode out to Table Mountain and will go there this P.M. to make observations. One point sought is proximity to the city that he may return nights after observing . . . Every honorable effort is being made to bring the observatory to Tucson. Dr. Comstock,[4] who is much interested . . . is unfortunately now in California . . .

[4]Theodore Bryant Comstock (1849-1915), a noted geologist and mining engineer, served as president of the fledgling University of Arizona during the years 1893 to 1895.

The *Citizen* went on to bemoan that the season was not the best. Douglass obtained his best results on Sentinel Peak ("A" Mtn) and Turtleback (Tumamok) Hill, low hills west of town, and even in the saddle between. But he was still unsatisfied with any of the spots visited, in part because weather conditions in the entire territory so far were less clear than had been anticipated. The Tucson city fathers were equally willing to donate land and offered other forms of assistance as an inducement for Lowell to establish his observatory in their midst, but Douglass had other places to test and boarded the train for Phoenix at midnight on the 16th.

Clouds and wind discouraged his efforts in the Phoenix area. On the Tempe Buttes, even when the wind had died down, the best at which he could rate the seeing was a six. In telegraphic communication with Lowell, back in Boston, Douglass expressed his dismay, suggesting that he be allowed to go further southwest and test the seeing at Yuma. He was told to wait out the weather a bit longer in Tempe and then go on to Prescott, as originally planned. After three more days, Douglass was able to resume his tests but achieved no better results and on 27 March took the stage to Prescott. Lacking space for the delicate telescope on the lurching stage, it was crated up and shipped east on the Southern Pacific to El Paso, thence north to Albuquerque and back west on the Atlantic & Pacific[5] and then south to Prescott via the newly completed Prescott & Arizona Central. Despite the roundabout trip of over a thousand miles, it arrived at Prescott almost coincidentally with Douglass.

Again, he was hampered by bad weather and could make no observations until the end of the month. Then, testing the seeing from some low hills south of the onetime territorial capital, he found only fair conditions and departed for Flagstaff on April 3rd. Blessed at last by good weather as the train brought him to his final destination, Douglass started his round of sightings in the high plateau area. Checking regularly with Lowell by telegraph, he made numerous tests from several locations, the most extensive series of his entire trip.

Having gone over the daily reports telegraphed to him, on April 16th Lowell replied that the point designated as "Site 11," a small hill just west of Flagstaff, seemed quite promising and on the same date wired instructions to begin preparations for an installation at that spot. The opposition

---

[5]This was soon to be amalgamated into the Atchison, Topeka & Santa Fe Railroad, of which Percival Lowell was a director. In this capacity, he had few compunctions about "pulling rank" and routinely arranged through its president, Edward Payson Ripley, to have limited express trains make special stops in Flagstaff for his convenience.

View of Flagstaff's Aspen Street toward "Site Eleven" in 1893 before it became known as Mars Hill. The Clark dome was to be in the exact center of the skyline.

of Mars was fast approaching and there was not much time to get the equipment set up and operating.[6] On 23 April, 1894, encouraged by the now usual promise of adequate land from the town and its leading citizens' commitment to construct a wagon road to the site, Douglass finally began to work.[7]

In the end the Observatory was given five acres of land by the town, and acquired a further seventy acres by purchase over the years that followed. Its major land-holding came fifteen years later from an Act of Congress which gave the Observatory use of a square mile to the west.

Construction proceeded apace, with Douglass engaging local laborers to prepare the sturdy mounting necessary for the telescopes Lowell and Pickering were arranging to have sent west any minute. Five weeks later, on 28 May, 1894, Percival was planning to arrive at Flagstaff, himself, and the first formal observations scheduled to begin.

[6]Due to the relatively high ellipticity of the Martian orbit, this was a particularly favorable opposition, not due to be repeated for seventeen years.
[7]The prime mover among the Flagstaff citizenry, in this as in much else, was the local lumber baron, Dennis Riordan.

Pickering's prefabricated 18-inch telescope dome being assembled in 1894

While Douglass had been testing the seeing throughout the Arizona Territory, W. H. Pickering was hard at work back in Massachusetts. He designed and built a dome to protect the instruments and then had it dismantled for shipment west; following in the wake of Douglass, it arrived in Tucson on 21 March. Pickering also arranged for the loan of an 18-inch refractor, only recently completed by the famous Pittsburgh lensmaker, John Alfred Brashear (1840-1920) and from his older brother's custody at Harvard, he was able to obtain the loan of a 12-inch refractor. Simultaneously the equally famed Boston astronomical instrument-maker, Alvan Graham Clark (1832-1897), made special attachments for these two telescopes so they could be mounted together, in parallel on one foundation, thus saving much time and substantial expense.

By 7 May Douglass's labor force had completed the footings and framework for the observatory's dome. Twenty-four cedar posts, set firmly into the ground of Mars Hill, were to hold the external housing, but the instruments needed something much more rigid to hold them precisely on target. A heavy iron tube was set deeply into a matrix of stone and wet concrete and, with Douglass checking its alignment with a transit, the final positioning was completed.

The two borrowed telescopes (12-inch and 18-inch) on their single mounting, 1894

He now had the problem of building the stationary walls of the dome and ensuring that the rotating portion (which had been reshipped north by the roundabout railroad route from Tucson) moved easily on its tracks. This required considerable adjustment; once the weight of Pickering's dome was in place it was found that the structure had to be dismounted and each of the eighteen beveled wheels (that ran in a track around the top of the stationary wall) reset before the dome would rotate freely.[8]

To help correct this malfunction, Douglass engaged the services of the Sykes brothers, English-born machinists, wranglers and operators of a local tinkering enterprise they called "Makers and Menders of Anything." (See Chapter 7) The elder brother, Godfrey (1861-1948), was afflicted with intermittent attacks of wanderlust that frequently called him to other ventures around the world, though he always returned to Arizona. However, Stanley (1865-1956) was to become a more permanent fixture around Mars Hill, designing and machining for forty-six years the delicate and unique gears, cogs and myriad other devices necessary to keep astronomical instruments correctly pointed in space while the parent Earth rotates in its daily and annual movements.

During this same period, the townsfolk of Flagstaff had been busy with their commitment and the promised wagon road was under construction. It ran up the draw north of the site and curved around to the crest of the hill overlooking the town. The friendly railroad had come through with the necessary telegraph poles for the observatory's essential telegraph connections and all was proceeding well. Nevertheless, back in Boston Percival was disturbed by cost overruns and cautioned Douglass to keep the total expense from becoming more astronomical than the observatory itself.[9]

Pickering showed up in Flagstaff on May 20th with the larger Brashear lens and it was soon mounted in the newly finished dome so as to be ready for use upon the arrival, one week later, of the man who was paying for it all.

From the end of May 1894 throughout the long summer of Arizona's high plateau, Lowell, Pickering and Douglass were almost constantly at the telescopes, making a wide variety of sketches and measurements of changes seen in the Martian features. Included were almost a thousand drawings of the planet's surface and over seven hundred determinations of irregularities along the red planet's day-night terminator line. It was a busy and productive summer and it laid the foundation for the colorful

---

[8] This was a predecessor to the historic dome referred to in Chapter 7.
[9] Lowell's exact words were ". . . lest they distance not the ephemeris."

and controversial dialogue that was to ensue between those who agreed with Schiaparelli, and now Lowell, about the "canali," and those who failed to see them at all.

Astronomers work strange hours; like owls and raccoons they tend to become semi-nocturnal creatures. Thus, very soon after commencing their intense nighttime exercises, the three observers were forced to reconsider their living arrangements. The Bank Hotel in Flagstaff was over a mile distant from their telescopes; they were routinely arriving on foot, tired and sleepy from a night's work, just when everyone else was bustling about the day's normal activities. This abnormal living pattern caused strains all around and soon Lowell authorized the construction of a cottage close by the dome, so that the astronomers could keep their essential lifestyle without interruption from the rest of the world. It was the first of the two dozen residences and support buildings that have since come to decorate the sylvan tranquility of Mars Hill (See Chapter 8).

Percival Lowell summarized the first year's work at his observatory in the preface to the first issue of the *Annals of Lowell Observatory*:

In 1894, observations of Mars at this observatory began synchronously with the visibility of the phenomena to be observed; for the yearly round of change, which the observations themselves disclosed, had but just started in the case of the southern hemisphere, the one chiefly shown us at that opposition, when the first observation was made. The chronological order of observation was thus the chronological appearance. The dual chronology has been kept in the presentation of these annals.

Underlying all the particular phenomena presented to us by the planet's disk is the fundamental one of change. The appearance of the planet does not remain the same day after day, or even hour after hour. Still less does the disk present a like appearance month after month. It changes, and changes in a way to show that the changes do not depend, except in certain instances, upon our point of view. The surface itself is subject to change and of a most striking kind, and upon this change as a thread are strung its particular manifestations.

Fundamental to the production of the phenomena, the fact of change is no less fundamental to an understanding of what those phenomena betoken. If we would read their meaning right, it is to the thread of causation that we must hold, in our distant journey of exploration. It thus becomes evident why the order of observations herein recorded is to the point. It was synchronous with the evolution itself.

Following the thread just pointed out, we shall, after the General Notes, take up successively:—

The South Polar Cap;
Changes Observed on the Surface of the Planet;
Canals;
Oases;
Canals in the Dark Regions;
Terminator Observations.
Change, although not specified except once in the titles of the papers, affects them all.
Particular studies not connected with seasonal development, such as those upon Martian Longitudes, the Micrometric Measures of the Diameters, and the Satellites, have been introduced in what seemed suitable places.

The *General Notes*, which followed the preface, dealt with the ground rules of the remaining report, its assumptions, credits, abbreviations:

On May 22, 1894, the first regular observation was made upon the planet; and on April 3, 1895, the last. In the interval it was observed on almost every night through June, July, August, September, October, November, and the first half of December, after which bad weather made the observations intermittent, until April 3, 1895, they were brought to a close. The planet [Mars] was thus kept in view practically continuously for a period of six months and a half, and with intermissions for nearly four months more. During this time were made: 917 drawings of its disk, in whole or in part, by Prof. W. H. Pickering, A. E. Douglass, and the writer; 57 measures of the snow-cap by Professor Pickering and Mr. Douglass; 13 polariscopic observations by Professor Pickering; 464 micrometric measures of the diameters, 403 by Mr. Douglass, 50 by Professor Pickering, and 11 by the writer; 79 observations on Martian longitudes, and 45 on Martian latitudes, both by the writer; 736 determinations of irregularities on the terminator by Mr. Douglass, Professor Pickering, and the writer, together with observations on both the satellites, and a search by Mr. Douglass for possible others.

During the period covered by the observations, the conditions relatively to the Earth were as follows, the data being taken from Marth's physical ephemeris of the planet,[10] with such alterations only as the observations themselves disclosed:
Apparent diameter, May 22   8'.4"
Oct 13 . . . planet's nearest approach to . . . Earth 22'.1"
April 3   5'.6"
. . . The instrument used was almost without exception the eighteen-inch, the

[10]This was the principal product of the obscure British mathematician and astonomer, Albert Marth (1828-1897).

powers employed being for visual purposes chiefly 440 and 617, and for micrometric ones 862 and 1305. In their methods of observing the observers differed. Personally the writer found a screen in the shape of a thin piece of ochre glass placed in front of the eye-piece as a rule conducive to detection of detail. He has since learned that Schiaparelli made use of the same device. Mr. Douglass preferred to observe without such screen. Professor Pickering held a middle position in the matter. Sunlight illuminating the field of view brought about similarly good results; as during June, when the planet was observed in the early morning, details came out best about three quarters of an hour after sunrise, although the seeing at that time was on the whole less good than earlier.

The quality of the air had a distinct effect upon the appearance of details. When the air was not steady the more difficult details, such as the canals, showed as broadish streaks smooching the disk. In the best air they contracted in width, standing revealed as narrower lines. The same thing was true of the round dots at their intersections . . .

3

Simply Call It The Lowell Observatory[1]
1895 – 1901

FTER THAT first season of operation on Mars Hill, Percival Lowell
remained unsure of the suitabil-
ity of Flagstaff as an observatory site, even though he had made many
seemingly satisfactory observations there himself. He continued to consider alternatives, either for a permanent location, or for temporary
observing opportunities. Thus, in early 1895 he ordered the dismantling
of the two borrowed telescopes; the one to be sent back to Harvard and
the other to Brashear.

Since Douglass still had much work to do by way of refining the raw
data from his earlier Arequipa observations, Lowell worked out a schedule based on Douglass's availability for him to perform a series of viewing tests at selected points southward through Mexico to the Equator.
Lowell himself would continue the optimization process by testing the
seeing along the westerly fringe of the Sahara Desert on selected high
points of Morocco's Atlas Mountains. In the end, however, nothing came
of the African alternative and the Mexican trip was aborted; Pickering's
dome stayed in place on Mars Hill, though temporarily empty.

However, in early July of 1896, Mars Hill came to life again with the
arrival of Lowell's own refracting telescope, a $20,000 instrument for
which the 24-inch objective lens had been made to his order from a blank

---

[1] This title is taken from a telegram sent by Percival Lowell to Andrew Douglass on 15
March, 1894, while the latter was in Tucson. The full message text was: "In answer to
name simply call it the Lowell Observatory."

The Bank Hotel in Flagstaff in 1892 with the Grand Canyon stage in the foreground

of glass that he had personally selected in Paris at the workshop of the famous French glassmaker, Edouard Mantois. This fine piece of lens-making craftsmanship was the first major telescope in the American Southwest and became the cornerstone of Lowell Observatory for two generations and, though later designated as a National Historic Landmark, remains in use to this day.

Arriving on the same train with the telescope was its maker, Alvan G. Clark,[2] and one of his daughters, along with Percival Lowell and his secretary, Wrexie Leonard. By the end of the month, with the help of some new astronomical assistants, W. A. Cogshall,[3] Daniel Drew[4] and Thomas J. J. See (about whom more below), Douglass and Lowell had the precious telescope installed and a program of regular observations underway.

But Lowell was still uncertain; the unusually monsoonal weather in

[2]Clark's wife had died in 1887 and his only son never reached maturity. The firm continued after his death, managed initially by its principal lens-maker, C. A. R. Lundin, and made two more telescopes for Lowell Observatory - its first 40" reflector and the 13" "Pluto" telescope.
[3]Wilbur Adelman Cogshall (1874-1951) spent most of his subsequent career in association with Kirkwood Observatory and the Indiana University Department of Astronomy.
[4]Daniel Abbott Drew, then thirty years of age, did some preliminary computing for Planet "X" and a bit of observing, but left Lowell's employ in June 1897 for a job as principal of schools in Baraboo, Wisconsin.

Sykes' 24-inch telescope dome assembled in Tacubaya, Mexico, in 1896

the summer of 1896 discouraged him about Flagstaff as a permanent location and another opposition of Mars would soon be at hand. In September he ordered Douglass to revive his studies of a Mexican alternative and contracted with the Sykes brothers of Flagstaff to fabricate a larger dome that would more comfortably house the larger telescope, but to have it demountable so it could be dismantled and erected elsewhere.

Because Mars was low in the southern sky for the 1896 opposition, it was advantageous to observe it from a lower latitude. Therefore, with help from Felipe Valle, the director of the Mexican Osservatorio Nacional, Douglass determined a spot on the high, central plateau near Tacubaya not far outside Mexico City and close to the location of Valle's own station. The prized lens had been shipped east in early November, when its owner carried it back with him to Boston; operations in Flagstaff were again suspended. Ten days later, on the 21st, the newly completed Sykes dome and associated gear were loaded on a freight car for shipment on the roundabout railway route to the south. By early December of 1896 Douglass, Godfrey Sykes, and a large force of local laborers were hard at work preparing the facilities at the Tacubaya site and, when Lowell arrived with the lens on the 28th, the final installation was sufficiently swift that observations began two days later.

While the resulting Mexican observations were satisfactory, the Earth soon orbited beyond the point of most favorable viewing for Mars and the staff then turned their attention to Jupiter and its satellites. Despite a slightly higher altitude, Lowell decided that, on the whole, conditions in Mexico were not all that different from Flagstaff and he might as well return to a more politically stable, congenial and familiar environment. Leaving his staff to wrap up observational details and then dismantle the equipment, in late winter Lowell returned home to Boston.

Douglass, however, an avid alpinist and soon to number among the founders of the American Alpine Club, took the opportunity of his presence near the fifth highest peak of North America to assay a climb. When his employer, who was sympathetic to the urge,[5] learned of his assistant's successful ascent of Popocatépetl, he wired his congratulations. At an annual salary of $800, though well above the yearly earnings of American factory workers, Douglass was hardly overpaid and Lowell could afford to be a bit liberal in regard to days off.

By the end of April 1897 the somewhat portable dome and its irreplaceable contents were on their way back to Arizona, where Godfrey and Stanley Sykes reassembled everything again on Mars Hill. The

[5]Lowell made several climbs in the Alps and was to die in office as president of the Appalachian Mountain Club, the oldest such organization in the western hemisphere.

Mexican dome porter winding clock weights at Tacubaya, 1896

Mars Hill in 1897 with the Sykes (Clark) dome in place

observing staff, consisting of Cogshall, Douglass and See, returned to Flagstaff on May 8th, and reinstalled the prized lens. But then Douglass was suddenly taken ill and the mantle of daily operating responsibility at the Observatory rested on the overly eager shoulders of Dr. Thomas Jefferson Jackson See (1866-1962), soon to become one of the more controversial figures in American astronomy.

Not only was Douglass out of action for several months, recuperating at his aunt's home in San Diego, but of even greater significance for the Observatory, back in Boston Percival Lowell was stricken with severe nervous exhaustion.[6] For most practical purposes he was to be out of action for the next four years and the burden of overall decision-making about the future of his observatory fell largely to Percival's third cousin, lawyer and brother-in-law, William Lowell Putnam, II.

Nine years earlier, Percival's younger sister, Elizabeth, had married this agnate grandson of George Putnam (1807-1878), the pastor of Roxbury's venerable First Church, an Overseer of Harvard and sometime member of the Massachusetts General Court. William's mother was born Harriet Lowell (1836-1920), a great granddaughter of the "Old Judge" by his third marriage (to Rebecca Russell Tyng). Despite its founder's illness,

[6]This seems to have been a fashionable affliction among many prominent Americans of that era - even including President Theodore Roosevelt.

Douglass sketching at the 24-inch dome in 1898

control of Percival Lowell's observatory remained very much in the family.

T. J. J. See had obtained his doctoral degree in 1892 from the University of Berlin and worked for the next three years under the eminent astronomer, George Ellery Hale (1868-1938), at the University of Chicago. He came with excellent academic credentials, but in time succeeded in alienating all those with whom he was working, particularly Douglass. Others in the field had noted See's apparent penchant for usurping the work of others and taking undeserved credit unto himself.[7] But being away from Mars Hill for several months, Douglass had not been personally subjected to the problem. However, See's abrasive personality soon prompted Cogshall and Drew to resign their positions and only when See left Flagstaff in November 1897 for an extended visit in the East, was Putnam able to induce Douglass to return from San Diego and assume charge of the Observatory.

See's return in the spring brought on a more direct conflict with Douglass and resulted in a series of letters from the remaining staff members to Putnam, in Boston, detailing the arrogance, bad manners and professional discourtesy that seemed to be his routine. By mid-July when hardly any staff was left on Mars Hill, the pot boiled over and Putnam fired See.

The totality of See's unfitness had been summed up in a lengthy letter from Douglass to Putnam under date of 28 June, 1898:

On receipt of your letter I consulted with Dr. Manning[8] about his sending you a written statement but as you asked for facts he seemed to think it unnecessary . . .

The doctor's opinion divided itself into two parts:—first, the existence of secret excesses, which if continued at See's age (31 years) are almost incurable, leading in the end to insanity or idiocy; and, second, actual evidence of mental degeneration which is incurable and gives positive warning of worse to come. These evidences include moral obliquity, loss of sense of right and wrong and cowardice, as well as mental peculiarities.

Cogshall and Waterbury[9] will tell you about the first part, the evidence of the

[7]More can doubtless be learned from William Larkin Webb's 298 page opus entitled, *Brief Biography and Popular Account of the Unparalleled Discoveries of Thomas Jefferson Jackson See;* Lynn, Mass., 1913.

[8]Felix Manning was the county health officer and maintained a private medical practice as well. His brother, Thomas, also of Flagstaff, became a prominent eye surgeon in Los Angeles.

[9]George A. Waterbury, Jr., worked at Lowell Observatory after 14 December, 1897, for almost two years. He was paid $25 per month plus board - but given the option to take $45 in salary but supply his own food.

Cogshall (left) and See on observing ladder of the 24-inch telescope, 1897

existence of such habits. Dr. Manning said that they had many times more evidence than are necessary to show the presence of that perverted sexual passion that follows intense sexual excess however accomplished.

Upon the signs of mental unsoundness or degeneration, the doctor's second point, I wish to speak more fully. I will mention several points in the order in which they occur to me . . .

Douglass went on to detail in three more pages of his typed missive, the evidence that he and others had observed regarding See's behavior under seven categories: Absent mindedness; Forgetfulness; Nervousness; Cowardice; Egotism; Desire for notoriety; and Jealousy. Then, gathering steam, he launched into a two-page recitation of See's unfitness to remain because of his "Treatment of subordinates and other people. Derision of Lowell's work and that of others."

Finally, his letter—totalling twelve legal-length pages—got to the point of greatest issue:

Third: Taking other people's ideas as his own. I have written much on this subject in previous letters. He has written four articles for astronomical magazines upon atmospheric currents, twinkling, &c, in which only two ideas are original with him. All the rest came from my conversations or publications, or the publications of others, and his articles were published during my absence from Flagstaff in the summer of 1897 . . .

He has heard my statement of discoveries and within two weeks told them to me as his own. I wrote this in a former letter. I was dumfounded but some incident then occurred which put it out of my mind for a time. After that I began to watch him. I have seen many less striking, or perhaps merely less suspicious cases. He is now writing on a new book and I know that he has copied many pages from G. H. Darwin's works . . .[10]

Douglass's recitation of See's iniquities ended in a postscript that deserves a place in some lexicon of evils:

Upon reading the entire above article to Dr. Manning his criticism was that I had not given enough importance to See's vice. Dr. Manning says that if he had an Observatory and See's ability were equal to a hundred Herschels and La Places combined, he would remove See at once. See's vice taints every act and

[10]Sir George Howard Darwin (1845-1912), one of the four prominent sons of Charles Darwin, was a mathematician, astronomer, authority on tidal friction, geodesy and dynamic meteorology.

thought and his presence defiles the young men at the Observatory. They cannot much longer refrain from serious trouble—fights by fists instead of words—and retain their self-respect. The Doctor says that words cannot describe the vileness of See's presence, the harm he is doing those in contact with him, and the promptness with which he ought to be removed.

Personally I have never had such an aversion to a man or beast or reptile or anything disgusting as I have had to him. The moment he leaves town will be one of vast and intense relief and I never want to see him again under any circumstances. If he comes back, I will have him kicked out of town.

See fulminated about his dismissal, and Douglass did him no favors in trying to find another job, but eventually he landed with the U.S. Navy at Mare Island in San Francisco Bay obtaining employment as a mathematician. By 1913 See held the rank of captain and had made his way into *Who's Who*, where he enjoyed a lengthy entry, which Douglass would probably have averred was largely undeserved.[11]

It was during this period that Douglass, frustrated with the reluctant movement action of the bigger dome, designed an arrangement for floating its rotating upper segment. This consisted of an elevated trough, filled with water, into which the upper portion was immersed to its flotation level. Putnam authorized a total sum of $600 for the project and it was constructed. Unfortunately, it had not been foreseen that cold weather might bring everything to a standstill. Furthermore, high winds caused the boat to rock, sloshing water down the lower sides of the building.[12] Before long the dome was put back on its original steel rollers. But in 1957, with money given by V. M. Slipher after his retirement, an array of 6" by 16" automobile wheels was attached to the dome so that it could roll gently around on the lower, fixed track, an unique concept which continues in operation to this date.

With See out of his way, and now designated by Putnam as "Assistant in Charge," Douglass continued his observational work and also ran Lowell's activities locally, making observations of various planets, improving the facilities on Mars Hill and continuing to contemplate alternative viewing locations. In this latter process he found Lowell unenthusiastic, perhaps apathetic because of his continued illness. Douglass thus compiled the bulk of the first two issues of the *Annals of Lowell Observatory* and handled almost all the scientific correspondence, includ-

[11]In the 1944 edition of WHO'S WHO, See's entry occupied more space than those for Albert Einstein, Edwin Hubble and V. M. Slipher - all combined.
[12]One can still make out the water stains on the lower dome woodwork.

The Clark dome complete in 1900. Note the wooden base and meteorological "beehive."

ing periodic defenses of his employer's views on the possibilities of intelligent life having once existed on the planet Mars. This occasional task, however, had started him thinking about Percival Lowell's scientific methodology and the verification of the lines that he and others were continuing to observe on the red planet's surface.

Percival Lowell was described by his sister, Elizabeth, as "a brilliant man . . ." And like many others of such capability, he was often impatient with the tedious processes by which verification of fact is necessarily determined in the sciences. His mind worked more like that of an artist, tending to leap ahead of tiresome work and redundant drudgery to the glory of a conclusion.[13] Douglass began to wonder if this trait made for sound discoveries—even worse, he began to wonder out loud.

An event in the spring of 1901, when Percival Lowell finally returned to Mars Hill, was critical. The year before, Douglass had made use of small globes, placed on the ground at appropriate distances from the telescope, to determine if astronomers might just be the victims of optical illusions in visualizing the markings that many had reported on the sur-

[13]Regardless of Lowell's scientific methodology - or lack of same - his ebullient pronouncements stimulated a lot of thinking and research by others that might never have occurred without his impetus.

Pontoons being assembled in 1899 for floating the 24-inch dome

face of Mars and more recently, Venus. Lowell had ordered this line of inquiry terminated, according to Douglass because it might cast doubt on some of the "findings" that had been published with such authority and fanfare from Lowell Observatory.

In the fall of 1900, however, Douglass was allowed to proceed again, and soon accumulated sufficient evidence to conclude that much of the detail previously reported might well be imaginary. Then came that fateful spring day when Percival Lowell, himself, observed one of these globes through a telescope and sketched a double line where only one existed on the target disc.

Douglass had consulted with an eminent psychologist about the possibility of the human mind working overtime on the observational process, seeing things that did not exist, imagining the seeing of detail already described by others, and being influenced by varying intensities of light at different times of observation. In all this line of questioning he was moving onto thinner and thinner ice with his employer. Douglass capped his uncertainties in a long and highly personal letter to Putnam under date of 12 March, 1901:

The object of my writing to you is this: will you use your influence with Mr. Lowell to make his work more scientific in character? My whole intimate acquaintance with his methods of thought and work, his Venus observations and articles and now recently his "Mars on Glacial Epochs" show that his work is literary and not scientific. In view of his approaching resumption of observational work it seems to me most important that everything possible should be done to preserve his scientific reputation.

It will be perhaps a year after we begin work on this new spectroscope that we will know enough about it to begin to publish, but I fear he will wish to publish results long before that. His work is not credited among astronomers because he devotes his energy to hunting up a few facts in support of some speculation instead of perseveringly hunting innumerable facts and then limiting himself to publishing the unavoidable conclusions, as all scientists of good standing do, in whatever line of work they may be engaged . . .

Douglass went on to cite a series of hastily written and careless publications, concluding his analysis:

The summary of this is that he is literary and not scientific. If he would confine himself to popular writing there is no limit to the greatness of his reputation as such a writer, but his method is not the scientific method and much if not nearly all that he has written has done him harm rather than good . . .

Now as to what can be done, I do not think Mr. Lowell would permit any minute supervision or rather criticism of his writings. But there is one thing that you can do and that is to urge him to go slowly with the real scientific writing, as they do in all other observatories of any repute,—to have the publications worked at all the time but never pushed . . . If, further, you could do anything to make him give his attention entirely to popular writing, you would advance his reputation in the best way. I fear it will not be possible to turn him into a scientific man. I have known for years that his methods were not scientific but I expected when Belopolsky[14] contradicted his results on Venus that he would see the necessity of using the scientific method. But I cannot see that he has done any good in his "Mars on Glacial Epochs" and I have lost hope. Lest you think that I have not thought out this matter thoroughly I will mention the most important defects I have noted in that last publication...

Douglass filled in the details and then played his strongest card in the next lines, asking that Putnam raise the questions of meteorological

[14]Aristarchus Belopolsky (1854-1934) was director of the Pulkova Observatory near St. Petersburg. He was a specialist in celestial mechanics and during this period did his observing with a 30-inch Clark refractor which had more than half again the light-gathering power of Lowell's.

veracity with Percival's first cousin, A. L. Rotch,[15] one of the leaders of that developing science and proprietor of the already notable Blue Hill Observatory south of Boston.[16]

> To return from this digression and consider what is to be done; there is a special case that will come up before very long. The one journal in the world where our spectroscopic work should be published is the Astrophysical Journal of the Univ. of Chicago, edited by Prof. Geo. E. Hale, Director of the Yerkes Observatory. But in 1895 Hale wrote Mr. Lowell that he could not publish any other contributions from him. Now what is to be done about that? He [Hale] made no such remark to me, yet out of loyalty to Mr. Lowell I have never sent anything there since. So we are barred out of the highest standard magazine in the world on that subject.[17]
>
> After months of careful consideration the situation appears to me to be sufficiently serious to warrant sending this kind of letter to you in order to obtain your help towards Mr. Lowell's best interests. For his real interests consist in his making a reputation that will be permanent . . .
>
> As for me, I have always been intensely loyal to Mr. Lowell and have served him faithfully. Seven hours a day is the time required at Harvard Observatory after 50 years experience because they consider that scientific work beyond that is poorer in quality and so does not pay. I have committed the mistake or questionable procedure of giving Mr. Lowell more than that and sometimes very much more, as for example during the years work on Volume II. But apart from that I am deeply attached to Mr. Lowell and would like to see his name have the highest and best renown, as that of the founder of the Lowell Institute does. It shames and pains me when scientific men say things derogatory to him. For example, a Frenchman in town here was written to by the Prof. of Anthropology in the Bruxelles Univ. for some Indian relics. The Prof. in the letter added that Flagstaff was well known there on account of its famous "farceur,"[18] Mr. Lowell, whose book on Mars he had read.
>
> Now I have presented the case as it appears to me. I give it to you fully because

[15] Abbott Lawrence Rotch (1861-1912), with maternal grandparents Abbott Lawrence and Katherine Bigelow in common with Percival Lowell, was also an alpinist of distinction and was to share an adventurous balloon ascent over London with Percival and his bride in 1908.

[16] For more on this observatory see *The Blue Hill Meteorological Observatory: The First 100 Years—1885-1985* by John H. Conover, published by the American Meteorological Society, Boston, 1990.

[17] The device used in this work, as amended and improved over the next several years, was the instrument with which V. M. Slipher was able to determine the enormous radial velocities of distant nebulae - discussed more fully in the next chapter.

[18] A contemporary colloquial translation would be "fabricator" or "joker."

Percival Lowell on the lecture circuit in 1896

William Lowell Putnam, II, with his wife, nee Elizabeth Lowell, at the family home in Manchester, Mass., 1902

you are in the family, as it were. (I have never given it to outsiders.) I would be very glad to have you verify what I have said as to the opinion of scientific men. More than that, you could do no one thing that would give me more pleasure and hearty relief of mind than to get a direct statement from some professional astronomer of international reputation, that I am wrong in this estimate and that Mr. Lowell's methods of scientific research are entirely correct and satisfactory. If you will do that I will give you my most hearty thanks and consider the matter settled in Mr. Lowell's favor. But that may be difficult, for out of the ten astronomers of greatest distinction in this country I have heard five—and some of those were Mr. Lowell's intimate and cordial friends—express opinions to the contrary, coupled in some cases, I can add, to kindly expressions of satisfaction that the outlay of money was put to some scientific use by my presence here.

I would be glad, Mr. Putnam, to have your kindly advice and assistance in this matter.

Very sincerely yours,
A. E. Douglass
Please consider this letter as between ourselves only.

While Putnam sent Douglass a cordial acknowledgment of this well meant missive, three months later, the "Assistant in Charge" was out of a job. His next few years were somewhat rocky, but ultimately settled into a most rewarding and productive career. After a fling at local politics he finally found employment in his chosen profession at the University of Arizona. In 1920 Douglass came to head the University of Arizona's Steward Observatory in Tucson, the second such institution to be established in the state, and in time to exceed Lowell's in size of staff and facilities.[19]

In 1901 while still working in Flagstaff—then and later a center of logging activity—Douglass had been struck by the possibility that careful analysis of tree rings might furnish a long-term weather report. He was hopeful of determining if the record of sunspot cycles compiled from visual observation could be carried backwards in time through comparative analyses of annual growth rings. First working with carefully polished log sections and then using a sophisticated coring device, he made microscopic analyses of the rings in the yellow (Ponderosa) pines prevalent across the high plateau of northern Arizona. He soon began to recog-

[19]Douglass worked hard to find the funding and establish the Department of Astronomy at the University of Arizona. Besides its own facilities, the nearby National Observatory and that on Kitt Peak have made Tucson more of a center for astronomy than Flagstaff. Maybe the weather really was unusual in March 1894 when Douglass first tested the seeing.

nize a consistent pattern that showed good growing years and bad, the wet and the dry, the warm and the cold. Further careful work showed that these patterns could be traced from the inner rings of some trees to the outer rings of others—often linking periods of hundreds of years, from tree to tree and into old wooden beams.

While Douglass's initial expectations from his science of dendrochronology seem only partially borne out by the subsequently noted record, even from those of the eight-thousand-year-old bristlecone pines, the process has become of enormous use to historians in dating human artifacts and constructions. The laboratory he established for this study—another part of the University of Arizona—is the world's acknowledged repository of learning on this branch of history. At the time of his death, at age ninety-four, Andrew Ellicott Douglass was both beloved and honored, not only for the quality of his work in two sciences, but for his modesty and low key manners in great accomplishment.

Undeterred by the doubts seeping into the mind of his departed subordinate, Lowell rebuilt his scientific staff around two Indiana farm boys, the Slipher brothers, Vesto Melvin (1875-1959) and Earl Charles (1883-1964), and the Norwegian-American, Carl Otto Lampland (1873-1951),[20] each of whose labors was to bring Lowell Observatory great distinction in the years ahead. Lowell's faith in their scientific prowess turned out to be remarkably well placed and they had the tact to express whatever feelings they may have held about their employer's personal scientific procedures in terms more acceptable to his flamboyant nature.

[20]Lampland's meticulously kept personal diary has been a valuable reference for this book. Interestingly, whenever he touched on a topic pejorative to either of the Sliphers, his notes took the form of Runic script.

4

Red Shifts and Gold Medals[1]
1901 – 1954

O NE FRIGID Flagstaff evening
at the very end of December
1912 Vesto Melvin Slipher made
preparations for an observation he knew would tax his own skills as well
as the capabilities of his instruments. He planned, with the aid of a spectrograph attached to the 24-inch telescope, to secure a photograph of the
spectrum of the Andromeda Nebula. This nebula—a leading example of a
type known as "spiral nebulae"—was, almost all astronomers judged, a
solar system in formation, a single star around which clouds of material
swirled. The nebula was perhaps a few tens of light years from Earth.

But to obtain a good photograph of the spectrum—a spectrogram—of
the Andromeda Nebula, Slipher knew would take many hours. He
would have to take great care throughout the exposure to ensure the
images stayed in focus and the telescope accurately pointed. In fact, to
accumulate enough light, Slipher extended the observation over three
nights, ending it only in the very early hours of 1 January, 1913. This
spectrogram, Slipher hoped, would be bright enough and clear enough
for him to measure the nebula's "radial velocity"—the speed at which it
was either approaching or moving away from the Earth.

If a light source is in motion with respect to an observer, then the
wavelengths of its spectral lines will move from the values they would
have in the absence of relative motion. The amount that the lines shift, the

[1] This chapter is the work of Robert W. Smith, who wishes to acknowledge the assistance
of Dudley Observatory in its preparation.

43

Doppler shift, reveals the relative motion between the source and the observer.[2] If the source is moving towards the observer the spectral lines are shifted up towards the blue, and if the light source is moving away, the spectral lines are shifted down towards the red—or redshifted. The first spectroscopic measurements of the radial velocity of a star had been made only in the late 1860s. By the 1910s, the radial velocities of over a thousand stars were known, but no nebula of the Andromeda type had yet had its radial velocity determined. As such nebulae were supposed to evolve into stars, astronomers expected them to move at about the same speed as stars. This expectation was shattered by Slipher's startling discovery that the Andromeda Nebula was moving towards the Sun at 300 kilometers per second, the highest speed then recorded for any astronomical body.

Soon there were more remarkable results coming from Mars Hill as Slipher turned his spectrograph and telescope to other nebulae. Within a couple of years, he had secured results indicating that almost all the spiral nebulae are fleeing the Sun at speeds of hundreds, and in some cases, thousands, of kilometers per second. Deeply puzzling to astronomers at the time, Slipher's results played a key role in a fundamental shift in astronomical theory. Within ten years of his original discovery, his findings, combined with some crucial notations by others, had convinced nearly all astronomers that objects like Andromeda were not really solar systems in the making, but huge collections of stars—galaxies. Within a few more years, astronomers generally agreed that the radial velocities of the spiral nebulae were to be interpreted as meaning we live in a continually expanding Universe.

Slipher is now regarded as one of the outstanding observational astrophysicists of the first half of the twentieth century and his results on the speeds of spiral nebulae are his greatest achievements. Who was the man who produced these startling results? What other finds did he make in his long career at Percival Lowell's observatory?

V. M.—as he was invariably known in later life—was born on a large family farm in Mulberry, Indiana, on 11 November, 1875, one of eleven children (two of whom died in infancy) of Daniel Clark and Hannah App Slipher. The family soon moved to Frankfort where V. M. received his early schooling.

He showed a ready aptitude for mathematics for a short time before

[2]Christian Johann Doppler (1803-1853), an Austrian physicist, first theorized on this topic in 1842.

The Slipher home in Mulberry, Indiana, 1920

entering Indiana University at Bloomington, he even taught the subject at a school near Frankfort. His parents were firm believers in the value of higher education and the choice of Bloomington was influenced by one of its graduates who had been among his teachers.

V. M. graduated from college in 1901 with a degree in mechanics and astronomy, only the third Indiana student to secure such a degree. The head of the department, and the person who gave him a love of astronomy, John Miller, was to become a lifelong friend and counsellor.[3] While in Bloomington, V. M. also met Emma Rosalie Munger, a fellow student who became his wife three years after their graduation.

Also on the astronomy faculty at Bloomington was Wilbur Cogshall, who had worked for Percival Lowell during 1896-7, and it was Cogshall who convinced a skeptical Lowell to hire the young Slipher. "As regards Mr. Slipher," Lowell wrote to Cogshall on 7 July, 1901,

I shall be happy to have him come when he is ready. I have decided, however, that I shall not want another permanent assistant and take him only because I

[3]John Anthony Miller (1859-1961) had a distinguished career in education and in astronomy. Later, president of Swarthmore College, he was also director of its famous Sproul Observatory.

John Miller (right) in discussion with Albert Einstein, 1929

promised to do so; and for the term suggested.[4] What it was escapes my memory. If, owing to this decision, he prefers not to come, let him please himself.

Despite these misgivings, Slipher arrived in Flagstaff a month later and Lowell promptly put him to work mastering a new spectrograph that had recently been ordered from John Brashear. This device is, in effect, a spectroscope with a camera attached to obtain photographs of the spectra. In the decade of the 1890s, the spectrograph, largely due to the increasing sensitivity of photographic plates, began to supplant the older spectroscope which required the astronomer to chart the spectral lines by eye. By 1901 and the arrival of Slipher and the Brashear instrument, the spectrograph was widely employed by astrophysicists. Slipher's ultimate mastery of its capabilities propelled him to the forefront of astronomy.

Percival Lowell's primary astronomical interest was the study of the solar system—for which he had built his observatory. The immediate cause of his decision to purchase a spectrograph was a solar system prob-

[4]Lowell had barely recovered from his lengthy illness and uncomfortable experiences with Douglass. This matter is referred to in Hoyt's *Lowell and Mars*, a very scholarly and invaluable resource in preparing this entire book.

lem—the rotation period of Venus. For Lowell, the stakes riding on this question were very high. He believed that on this question "... turns our whole knowledge of the planet's physical condition. More than this, it adds something which must be reckoned with in the framing of any cosmogony."[5] Moreover, Lowell had already claimed that markings (spokes) he had detected on the cloudy Venusian surface, and which were even more controversial than the Martian canals, disclosed that Venus rotated only once in each orbit around the Sun, in the same manner that the Earth's moon always keeps the same face towards the Earth. However, from any inclinations of its spectral lines one might deduce an independent estimate of its rotational period, one that Lowell hoped would support his own visual observations.[6]

Lowell was very conscious of the need for good instruments and had urged Brashear to produce the best spectrograph he could make. Thus, before completing its design, Brashear consulted some of the users of his existing astronomical spectrographs for their opinions on improvements that might be made in their own instruments. Brashear's company was an extremely important resource for American astronomers and astrophysicists in this period. He was someone to whom they could turn for instruments at least as good as those being manufactured in Europe. Trained as a machinist, Brashear had worked in the steel mills of Pittsburgh before setting up his own shop. Brashear, however, was not much of a businessman, nor did he have to be. William Thaw (1818-1889), a Pennsylvania transportation magnate and patron of Pittsburgh's Allegheny Observatory provided financial support for Brashear for many years.[7] As with so much of late nineteenth and early twentieth century American science—very well exemplified at Lowell Observatory—private patronage provided the financial muscle; in Brashear's case via the subsidy of one of the chief suppliers of instruments.

There was, however, no question of Percival Lowell purchasing an "off-the-shelf" instrument; Brashear, as he invariably did for his major orders, had to produce a custom-built instrument, designed in collaboration with its eventual user. His shop was already experienced in their

[5] See P. Lowell, *The Evolution of Worlds*, Macmillan, 1909, page 760.
[6] This controversy continued long after Lowell's death. Half a century later, conclusive radar measurements finally showed that the rotation of Venus was indeed very slow - even retrograde - with a day equivalent of 116.75 Earth days and the Sun rising in the Venusian west.
[7] For more on Thaw, see Donald Osterbrock, *James E. Keeler, Pioneer American Astro-physicist*, Cambridge University Press, 1984. Thaw's wife, Mary, continued his scientific philanthropy for another thirty years after her husband's death.

construction and so he and his staff could draw on their earlier work, skills and physical plant, to design and build Lowell's spectrograph. In fact, the Lowell spectrograph is best seen as a modification of the Mills spectrograph that Brashear had delivered to the Lick Observatory in California in 1894, and which had been designed in outline by William Wallace Campbell (1862-1938), and in detail by Brashear.

This sort of relationship between an instrument maker and his client astronomers was not common during the nineteenth century. In Victorian Britain, for example, Astronomer Royal George Biddell Airy (1801-1892) regarded instrument makers as tradesmen and treated them accordingly.[8] With Brashear, the instrument maker was far more of a collaborator than a salesman. Hence when an astronomer secured a new instrument, Brashear and his workmen did not simply deliver the instrument but devoted much subsequent effort to making sure it worked.

After the Brashear spectrograph arrived on Mars Hill in 1901 there were numerous interchanges between Lowell, Slipher, and Brashear and his staff, interchanges of information, as well as prisms, screws, thermometers and other pieces of apparatus. These were designed to help Slipher implement Lowell's charge to put the spectrograph into perfect running order—to tame the new instrument.

At the turn of the century a large astronomical spectrograph attached to a sizeable telescope presented major challenges even to adroit and experienced astrophysicists. For example, Campbell trained under the outstanding spectroscopist James Edward Keeler at Lick Observatory. When Keeler left Lick, Campbell succeeded his position and assumed responsibility for Keeler's program of spectroscopy. When a new, large Brashear spectrograph arrived in 1894—the Mills spectrograph that Campbell had helped to design—Campbell, who was to become one of the leaders of American astronomy, immediately ran into serious difficulties.

When Slipher started to grapple with Lowell's Brashear spectrograph he was in a worse position than Campbell had been. First, despite his degree in mechanics and astronomy, he had relatively little practical astronomical training. When he arrived on Mars Hill the biggest telescope he had handled was a $4\frac{1}{2}$-inch reflector, a tiny instrument compared to the Observatory's 24-inch Clark refractor. Slipher had taken a course in spectroscopy at Indiana, but from his student notebook it is clear that he could have had only an elementary knowledge of spectral

[8] Airy's cavalier treatment of John Couch Adams' calculations was also of this nature. See page 173.

theory. Certainly he had never handled an astronomical spectrograph of the size and complexity that now confronted him.

Lowell's observatory was—then and now—a private research institution, so there were no teaching responsibilities for Slipher. He could devote himself to mastering the telescope and spectrograph. But, though lacking the distractions of teaching, Slipher was isolated astronomically in Northern Arizona, hundreds of miles from the nearest other observatory, professional astronomer or commercial instrument workshop. Percival Lowell was himself only a part-time visitor and had little knowledge of spectroscopy; Slipher thus had to start to master the spectrograph by a combination of correspondence course, learning by doing and some instruction from Brashear's staff.

Not surprisingly he got off to a rocky start. By December 1901 he was pleading to go to Lick to learn first hand how astronomers there employed the Mills spectrograph. Lowell hated the idea. His relations with Campbell, now the Lick director, were extremely bad, a consequence of the Mars furor. Lowell told Slipher he should think of visiting Lick only when he'd become thoroughly familiar with the spectrograph ". . . and can give them as much as you take."[9]

In March 1902 Slipher could still get the blue and red ends of the spectrum confused. By summer he was doing rather better and Lowell was pleased enough with spectrograms of Jupiter and Saturn that he sent copies and an enthusiastic letter to Brashear. Indeed, Slipher soon confirmed the well-established rotation periods of Jupiter, Saturn and Mars. He thereby demonstrated a degree of competence with the spectrograph, even if these measurements hardly posed a novel or particularly challenging problem. Lowell decided the next steps were for Slipher to work on two areas of key importance for his own solar system program. One was the rotation period of Venus, the problem which had prompted the acquisition of a spectrograph.

Five months after starting on the Venus problem, Slipher had evidence from the inclination of the planet's spectral lines that Lowell seized upon as supporting a "slow" rotation for the planet, justifying his claim, based on his observation of the "spokes" on the planet's surface, that Venus rotated only once in each orbit. To Lowell, Slipher's spectrograms were conclusive; indeed, he decided that it would be "unnecessary" to ever have them repeated, but Slipher had now really started to "deliver the goods," and though initially hired for a short term, had become a permanent fixture at Lowell's observatory.

[9]P. Lowell to V. M. Slipher, 28 December, 1901.

The Lowell Observatory staff at the 24-inch dome in 1905. From the left are Harry Hussey, Wrexie Leonard, V.M. Slipher, Percival Lowell, C.O. Lampland, and John Duncan.

Slipher was still, however, very much an assistant—his primary role to perform those tasks his employer set. Lowell, often in Boston, expected to be kept fully informed of what was happening at his observatory and maintained a steady stream of instructions and advice to Slipher, including such mundane details as caring for the Observatory's cow, Venus.

The flow of instructions did not mean Slipher was a mere cipher for Lowell was in no position to tell Slipher precisely what to do. He also allowed Slipher, and later other astronomers on the Observatory staff, latitude in pursuing his—that is, Lowell's—interests. Also, in the time not taken up by the pursuit of projects designated by Lowell, Slipher had a relatively free hand. Hence, Lowell Observatory, although unusual in the early twentieth century in being a private research observatory both run and financed by one individual, was not unusual in how its staff was directed. In fact, Slipher had probably more flexibility than if he had been based, say, at Lick, where the mentality was that of a "radial velocity factory," whose director's chief interest was in collecting the radial velocities of stars to ever fainter limits. Lick's operation was owed something to the attitude that dominated the activities of many professional scientists in the nineteenth century. It led to prodigious quantities of information collected, sometimes published in thick volumes of which the pages still remain uncut. While these massive data collection programs attracted some professional astrophysicists, they repelled Lowell. He directed his assistants to specific problems, but avoided production line methods and goals.

While Lowell's concern was with the solar system, he did allow Slipher to study objects outside its reaches. In fact, almost as soon as Slipher had the spectrograph "under control," he had begun to observe stars. He embarked, for example, on a program of observations of the so-called standard velocity stars, that is, stars selected by an international group of astrophysicists for detailed measurements of their radial velocities at several observatories. Here, by participating in this program, was an excellent means for Slipher to check his own skills and the qualities of his instruments against those of the leading practitioners in the field, as well as establishing his credentials in an area outside those with which Lowell Observatory was usually associated. Doing so marked another important step in his mastery of the spectrograph. For, as Slipher pressed his measurements, he became convinced, that, as he wrote, "The full power of the spectrograph has . . . not been realized and the agreement of the velocities from different lines of the same plate is not so close as it should be with a spectrograph of this size."[10]

This conclusion by V. M. marks the end of his apprenticeship with the spectrograph. He had tackled problems of major concern to astronomers and his findings had been published in major astronomical journals. In so doing, he had obtained what he believed was a good understanding of

[10]V. M. Slipher, "Observations of Standard Velocity Stars with the Lowell Spectrograph," *Astrophysical Journal*, XXII (1905), page 319.

Carl Lundin cleaning the objective lens of the 24-inch telescope, 1901

Percival Lowell with his cow, Venus, in 1905

his instruments, but was now convinced of its limitations and the need for modifications if he was to do much better.

By 1905 Slipher had begun looking to improve the spectrograph, a process which involved him in intense practical activity. He tried different kinds and numbers of prisms for dispersing light; different kinds of photographic plates and even dyes for those plates to record the spectra. In a patient, often cut and try fashion, he sought, in particular, to wring out of his instrument every last bit of response at the red end of the spectrum. In all this he was greatly aided by Stanley Sykes, who had joined the Observatory full time in 1910.

During his own research program on the stars, Slipher noticed in 1908-9 that a number of spectra of stars, including some double stars, exhibited lines he interpreted as due to the presence of clouds of calcium between the Earth and the star. In so doing, he supported the 1904 findings of the German astronomer Johannes Franz Hartmann, but whereas Hartmann had results for one star, Slipher secured results for many more. Although these findings won Slipher some acclaim at the time, the importance of these researches was more fully understood two decades later when the existence of huge quantities of gas between the stars became generally believed by astronomers.

Given Lowell's own priorities, however, the stars had to wait their turn behind the planets in Slipher's observing schedule. As early as 1903, Slipher had started spectrographic observations of Uranus, Neptune, Jupiter and Saturn. By 1905 he had a number of photographic plates which disclosed a variety of groups of dark spectral bands, many of which eluded his efforts to identify them. Indeed, it was not until 1931 that German astronomer Rupert Wildt convinced his colleagues that some of the bands might be due to ammonia and methane. Shortly after Wildt's identification, the young physicist Arthur Adel joined the Lowell staff. Aided by a first class training in the new quantum mechanics at the University of Michigan, Adel seized on Slipher's plates and during the late 1930s identified more of the bands in a string of papers, some coauthored with Slipher. It was an impressive—if reluctant—marriage of old and puzzling data with some of the latest tools in theoretical physics.

In 1906 Percival Lowell set Slipher to work studying the spectra of light from spiral nebulae, all of which were then generally thought to be within the Milky Way.[11] Lowell was interested in proto-solar systems, a line of thought that dovetailed neatly with his program of revealing the history and evolution of our solar system. He wanted Slipher to see if the spectra of the outer regions of spiral nebulae were similar to those of the spectra of the giant outer planets. At first, because the spirals are so faint, Slipher did not think much of his chances for success. Few spectrographic observations had been made of them before. Astrophysicists had preferred to work on the spectra of bright stars which were relatively easy targets; the nebulae, as their very name implied, were much more elusive.

But his employer was determined, and in 1909 Slipher geared up to attack the problem. He wrote to other astrophysicists for suggestions on how to proceed, as well as experimenting with various changes in the spectrograph. By November 1910 Slipher had taken the crucial step of convincing himself that the key to obtaining useful spectra of the spirals was the speed of the spectrograph's camera lens, not, for example, the extent to which his prisms could disperse light. Thus, the largely self-taught Slipher had thought out for himself the most important single factor in his task, a concept not to be gleaned from any astronomical manual of the period.

By this stage, he had also shifted to a single prism, rejecting the standard three-prism arrangement. With these modifications, fashioned largely by Stanley Sykes, he could cut down drastically on the exposure times

[11] For much more on the debate in the early 20th century about the nature of spiral nebulae, see Robert W. Smith, *The Expanding Universe: Astronomy's 'Great Debate' 1900-1931*; Cambridge University Press, 1982.

needed. In fact, Slipher estimated that he had improved the capability of the apparatus by a factor of 100 over that which his scientific peers were using. Deploying this version of the spectrograph, Slipher secured a spectrum of the Andromeda Nebula on which he claimed to Lowell that he could detect faint "peculiarities not commented upon" by earlier researchers in the field.[12] "These early observations were made with large reflecting telescopes," he continued, "and the idea seems to go undisputed that a long focus telescope and of course a refractor is unsuitable for such work. But I convinced myself that I knew of no reason why the focus to aperture ratio had the slightest part to play in the spectrum work on extended objects, and this plate proves the proposition to my mind."

Slipher was also very busy with other work for Lowell, but when time allowed during 1911 and 1912, he continued his experiments with the spectrograph and spiral nebulae. By September 1912 he was using a very fast, commercially available, camera lens that he estimated meant his latest version of the spectrograph was about 200 times faster than the usual three-prism device.

On 17 September he took a six-hour exposure of the Andromeda Nebula. This plate revealed sufficient detail for him to think of measuring the shifts of the Nebula's spectral lines in order to determine its radial velocity. In pointing this out, he was careful not to raise Lowell's hopes too high because the radial velocity of a spiral was a prize definitely worth winning. As Lick Observatory director Campbell put it in his Silliman Lectures at Yale in 1910, "There is certainly no more pressing need at present than for a greatly increased number of nebular radial velocities," and no one had yet measured even one spiral's velocity.

After the 17 September plate, Slipher followed up with two more plates spread over a total of four nights. These, although still very faint, showed even more detail, certainly more, Slipher judged, than other researchers had recorded. On 28 December he began a three-night exposure on the Andromeda Nebula. After checking this plate, he told Lowell that "the velocity bids to come out unusually high."

In the next two weeks he measured his four plates of the Nebula, arriving at the conclusion that it was moving towards the Sun at about 300 kms/sec. It implied, interpreting the spectral shifts as Doppler shifts, the Nebula was moving three times faster than any other object in the universe. This was a remarkable result, made even more striking by the fact that nebulae were generally expected by astrophysicists to have low velocities—comparable to and perhaps somewhat lower than the stars—

[12]In a letter of 3 December, 1910.

speeds at most of a few tens of kms per sec.

Was the result to be trusted? Slipher remeasured his plates and sent a copy of a spectrum to the Lick Observatory. For a time he wondered if the spectral shifts actually indicated a velocity shift. But with his remeasures complete and no dissent from Lick, he announced his find to Lowell, who responded: "It looks as if you have made a great discovery. Try some other spiral nebulae for confirmation."[13]

A very large spectral shift had been found earlier by Fath at Lick who had measured a globular star cluster's radial velocity.[14] But his value was so much larger than he had expected that he had dismissed it as a fault of his spectrograph. Fath, then a graduate student, knew what sort of result to expect and so discarded the apparently anomalous value. By 1912 Slipher was an experienced practitioner of spectrography and had developed a very strong feel for when his instrument was transmitting unreliable messages. Although his result for the Andromeda Nebula was anomalous within the context of the understanding of spiral nebulae, he never thought his instrument was at fault. He was also encouraged by enthusiastic words from his onetime teacher, John Miller, who wrote him. "It looks to me as though you have found a gold mine, and that by working carefully you can make a contribution that is as significant as the one that Kepler made, but in an entirely different way."[15]

By the time of the 1914 meeting of the American Astronomical Society, Slipher had secured more results. Some nebulae displayed even larger shifts than the Andromeda Nebula, mostly to the red and, interpreted as Doppler shifts, these gave large velocities of recession. These results— many based on plates exposed for tens of hours, which meant they had to be exposed over several nights—were widely regarded as a tour-de-force. They won Slipher a standing ovation from his colleagues when he announced results for fifteen spirals, including two spiral nebulae that were receding at the staggering speed of 1,100 kms per sec. Following Slipher's example, a couple of confirmatory results were secured by astrophysicists elsewhere. Now no one seriously questioned the reality of the spectral shifts that Slipher had measured. The novice astrophysicist who a little over a decade earlier had confused the red and blue ends of the spectrum had, in the company of his spectrograph, come a long way indeed. In the process, both Slipher and his spectrograph had been transformed.

[13] P. Lowell to V. M. Slipher, 18 February, 1913.
[14] Edward Arthur Fath (1880-1959), German-born, was director of Carlton College's Goodsell Observatory from 1926 to 1950.
[15] J. Miller to V. M. Slipher, June 1913.

Although he had been initially hesitant about interpreting the shifts of the spectral lines as Doppler shifts, he soon decided that the preponderance of redshifts really did indicate that the spirals are fleeing from the Milky Way. But when V. M. delivered his 1914 address he was still not prepared to concede that the spirals were groups of stars, let alone galaxies. A number of astronomers explained his results in a very different manner. The famous Danish astronomer, Ejnar Hertzsprung, sent Slipher his hearty congratulations on his "beautiful discovery" of the high radial velocities of the spiral nebulae. To Hertzsprung and others, the speeds of the spirals seemed altogether too great for them to be held gravitationally to the Milky Way system of stars. For these people, Slipher's results meant that the spiral nebulae are in fact galaxies of stars lying far beyond the Milky Way. By 1917 and with more measurements of the speeds of spirals, Slipher agreed with them.

In the late 1910s and early 1920s, the nature of the spiral nebulae was a hotly disputed topic. More and more astronomers came to accept the idea that spiral nebulae are indeed huge galaxies of stars. For many, the really clinching evidence was secured in 1923 and 1924, when Edwin Powell Hubble, of the Mount Wilson Observatory, detected so-called Cepheid variables in the Andromeda Nebula. The Cepheids enabled Hubble to calculate the Nebula's distance with what astronomers agreed was unprecedented accuracy. His answer was roughly a million light years, a value which put the Nebula far beyond the boundaries of the Milky Way.

A few years later Hubble took another extremely important step when he plotted the distances of the galaxies against their radial velocities. The resulting plot seemed to prove that there was a relationship between a galaxy's distance and its redshift: the more distant the galaxy, the greater the redshift. Moreover, the relationship was roughly linear, that is, each time the distance was doubled, then the radial velocity doubled too.

The velocity-distance relationship was soon interpreted by astronomers as the consequence of an expanding universe. Hence some writers refer to Hubble's discovery of the "expanding universe." But this simplistic attribution of credit for the discovery of one of the central tenets of modern astronomy misses two significant points. Hubble was not the first to plot velocities and distances of galaxies, although his initial plot (drawn in 1929) was the first that other astronomers found to be really convincing—and—almost all the radial velocities used for his plot had been secured by Slipher. Hubble later told Slipher that the velocity-distance relationship was the product of his distances and Slipher's radial velocities. In praising Slipher's results of 1912 and 1913 on the

Andromeda Nebula, Hubble said "I regard such first steps as by far the most important of all. Once the field is opened, others can follow."[16]

There were other important spectrographic findings to come, although none would match the importance of his results on spiral nebulae. Slipher's researches on the light of the aurorae and night sky, for example, were significant pioneering studies. Auroral displays are rarely seen from Flagstaff, but his spectrographic scrutinies of those that were visible revealed features interpreted as the products of nitrogen, sodium and the other elements in the Earth's upper atmosphere in various stages of ionization.

In examining Slipher's scientific papers on the spectra of the aurorae and night skies, as well as on various other astronomical objects, one gains a sense of his scientific style. He laid the emphasis squarely on securing reliable observational results and in so doing almost always tried to eschew speculation. This tendency became even more noticeable after the death of Percival Lowell in 1916.

Slipher had been made the assistant director the year before and upon his employer's death he became the Observatory's acting director, although he was not officially made director until 1926. The decade 1916-26, as seen elsewhere in this book, saw the bruising and, for the Observatory's finances and prospects, devastating battles over Lowell's will. Despite the best efforts of the Observatory's trustee, this was a time of deep and often disturbing uncertainty for the small staff subject to the capricious and at times bizarre acts of Lowell's widow, Constance. In this period Slipher began to develop interests outside of astronomy that he may very well have initiated as a hedge against Mrs. Lowell. He shrewdly bought ranch property around Flagstaff, ran a retail furniture store for a period, was a founder of Flagstaff's community hotel, the Monte Vista, and purchased rental properties around the area. By the end of his life he had become a rich man.

But Percival Lowell's death had an even more far reaching consequence for the Observatory than the fight over his will. The staff, Slipher included, hesitant to publish their findings even when Lowell was alive, now retreated to a large degree into their shells. The controversies stirred up by Lowell's claims about intelligent life on Mars had swirled around the Observatory for years. Lowell's theories played a key part, for example, in a controversy over Slipher's claimed detection of water vapor in the atmosphere of Mars. This claim was hotly (and it would turn out

[16]E. P. Hubble to V. M. Slipher, 6 March, 1953. See also Edwin Hubble, *The Realm of the Nebulae*, Yale University Press, 1936, page 105.

V.M. Slipher studying Triassic dinosaur footprint west of Tuba City in 1928

much later, correctly) disputed by W. W. Campbell. This argument exasperated the even tempered Slipher. In 1908 he confided to Miller that he had a distaste for controversy, and had decided that the Observatory had been so heavily criticized that "I have come to the conclusion that where we can defend ourselves—we shall have to do it otherwise everything we

Building the Monte Vista Hotel, 1926

publish will be discredited."

But with Lowell dead, Slipher and the other permanent members of staff—his brother E. C. and Carl Lampland, both of whom, like Slipher, had joined the Observatory from Indiana University in the 1900s—chose a deliberately conservative publication policy. In line with their own temperaments as well as in reaction to the Mars furor, Slipher and his colleagues decided that if they were not utterly sure of a result they would not publish it.

Slipher did not hero worship Lowell, as did his brother and Lampland, both of whom would remain totally convinced Lowellians throughout their careers. Nevertheless, even Slipher felt intense respect and admiration for Lowell and believed that the wider astronomical world had never properly appreciated him. Nor, during his astronomical career, did he doubt the existence of canals on Mars. Further, Slipher and his colleagues believed that Lowell had pointed out the research routes down which they should travel, and they intended to follow his directions.

One line of research inspired by Percival Lowell that was ardently pursued by the Sliphers and Lampland was the hunt for a planet beyond Neptune, which Lowell had referred to as Planet X. As described elsewhere in this book, the chase was crowned with success in 1930 when

Marcia Slipher's wedding party, August 10, 1931. From the left are: C.O. and Verna Lampland, Althea Jones, Mrs. E.C. Slipher, Capella Slipher, E.C. Slipher, Allan Cree, Marcia Slipher Nicholson, David Slipher, Kenneth Nicholson, Rector Johnson (minister) and Mr. and Mrs. V.M. Slipher.

Clyde Tombaugh detected a tiny spot of light that switched its position between two photographic plates exposed at different times. The spot of light turned out to be Pluto. But while Tombaugh had made the actual find, he had only done so because of a concerted and long-term team effort at Lowell Observatory led by Slipher. Here was a triumphant vindication for Slipher and his colleagues of their faith in a goal they had pursued for over twenty years.

In fact, the early 1930s provided many of the high points of Slipher's astronomical career. Not only did these years see the discovery of Pluto, but also the award of several very prestigious prizes and so the recognition of his talents and achievements by his astronomical peers. Among the most satisfying of the prizes for Slipher was the Royal Astronomical Society's Gold Medal, presented to him in London in 1933 for his work in planetary spectroscopy. As the president of the society announced in also praising his studies of the radial velocities of spiral nebulae:

In a series of studies of the radial velocities of these island galaxies he laid the foundations of the great structure of the expanding universe. . . If cosmologists

In 1936 V.M. and E.C. Slipher stand near the east entrance of the building later named in their honor.

today have to deal with a universe that is expanding in fact as well as in fancy, at a rate which offers them special difficulties, a great part of the blame must be borne by our member . . .

But if the early 1930s saw public and professional acclaim for Lowell Observatory and Slipher, the Great Depression made it a difficult period for the Observatory, as for many other institutions. Without the energies of the trustee, Roger Lowell Putnam, and a generosity which extended to delving into his own pocket, the Observatory would have been in much worse shape. As it was, there was even a small expansion of its staff in the 1930s—one of whom was Arthur Adel.

Adel's expertise has been referred to above. But Slipher, like others, was not especially in harmony with or in command of the newer sorts of astrophysics developed in the wake of the conceptual revolutions in physics. His own training, experiences and inclinations were those of an observer who had learned to be deeply skeptical of theory, and sometimes of theorists, too. The new breed of highly trained and often theoretically sophisticated astrophysicists also tended to lack the sort of deference that Slipher's generation had extended to its elders.

V.M. Slipher in discussion of spectrograms with Lampland in 1947

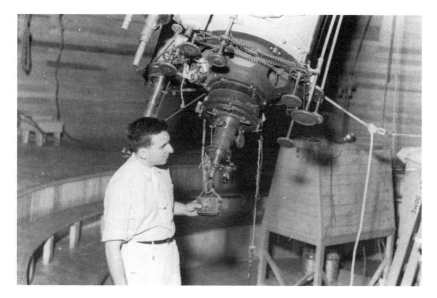

Arthur Adel at the 24-inch telescope, 1936

Slipher was more comfortable with the young Tombaugh—who had arrived at Lowell in 1929 with no formal training at all in astronomy and at the time even lacking a college degree—than with the more questioning and self-confident Adel. Perhaps in the Kansas farm boy, Slipher saw something of himself. It is significant, too, that Adel's presence at Lowell owed much to the urging of the trustee who saw a need for developing the staff with younger scientists.

Adel gave Lowell Observatory a dimension it had previously lacked. His appointment also pointed the way to a possible future, one in which the latest astrophysical theory and observation became much more tightly linked, as in his researches using Slipher's older plates. This, however, was not a road down which Slipher felt comfortable travelling. Also, for someone who had spent the first fifteen years of his professional career under Percival Lowell's directorship when money had been of little object, the financial stringencies of the later years were difficult. Nor did he develop successful strategies to secure substantial extra funds. But, in the 1930s, survival itself was no mean achievement.

By the end of World War II, the Observatory was barely surviving, almost no scientific papers were forthcoming, Adel and Tombaugh had left and relatively little use was being made of the several telescopes on

63 / *Red Shifts and Gold Medals*

Mars Hill. By then the key staff members were all over sixty years of age and the only professional environment they had ever known was Lowell Observatory. Slipher, himself, had become withdrawn from the profession. The Observatory was rescued from this perilous position by the foresight and determination of its trustee. As Mayor of Springfield, Massachusetts, he had proved adroit at bringing federal funds to his city; now he pressed Slipher to take advantage of the federal government's radically changed attitude towards the sponsorship of scientific research.

Most of the funds for American science before World War II came from industry, private foundations and state universities. During the war years this pattern changed dramatically with the federal government taking an ever-increasing lead in sponsoring scientific enterprise. Like many of his generation, Slipher was at first deeply skeptical of this process and wanted no part of it—what came one day, might disappear as rapidly the next. As a political conservative who had felt uneasy about the government largess of the New Deal, he feared being tied to the apron strings of distant and uncontrollable interests. But the trustee prevailed, convincing him that this was the wave of the future and Lowell Observatory could not expect to live in the past.

The influx of federal funds was to lead to a remaking of Lowell Observatory, though the estate of its founder continued to form a substantial part of its income and so provide a measure of independence. The first new endeavor was sponsored by the U S Weather Bureau—to study plan-

Far left: V.M. Slipher, Mary Russell Colton, Andrew Ellicott Douglass, Harold S. Colton (Director of the Museum of Northern Arizona), at V.M.'s retirement party in 1954

Left: Henry Giclas and incoming director Albert Wilson with gifts for the guests of honor

etary atmospheres through an external look at the circulation on Jupiter and Mars. This soon led to another long-term study (see Chapter 12) on solar variations and their impact on terrestrial weather patterns. The staff grew, the scientific productivity increased, new apparatus and methods arrived and, when Slipher retired as director in 1954, the Observatory's prospects were brighter than they had been for decades.

Percival Lowell was the despair of many professional astronomers, but the most scientifically productive period for V. M. Slipher was undoubtedly the decade after 1906, when he completed his mastery of the Brashear spectrograph and before his patron's death. Inspired and pressed by the Bostonian patrician, Slipher, the Indiana farm boy, had done great things. Through unflagging patience and his command of state-of-the-art instruments at a good observing site, Slipher had racked up a string of first-rate achievements. He was in the forefront of a generation of American astrophysicists that helped to confirm the shift from Europe to the United States of the center of observational astrophysics and cosmology.

Unfortunately, Lowell's late-in-life marriage and premature death caused problems to arise in the furtherance of his scientific image and seriously delayed the acceptance of his employees' work as being of the first rank.

II

An Unnecessary
Setback

Dr. and Mrs. Percival Lowell in their wedding portrait, 1908

5

Constance vs Wrexie
1893 – 1916

ASTRONOMERS ARE no more celibate than the rest of human society, though they tend to nocturnal diversions on a more cosmic scale than their fellows. The complete nature of Percival Lowell's love life is difficult to fathom at this distance in time, particularly in view of the deliberate complexities (and omissions) that seem to have been introduced into (or expunged from) the record over the years.

During his years as a Boston bachelor, Percy was quite the man about town, attending the usual rounds of parties and even becoming on one occasion the victim of an assault by a frustrated swain. In that well documented instance, it seems that in the autumn of 1886 the Walter Cabots held a soiree at their home to which many acquaintances were invited, among them one John Jay Chapman, then aged twenty-five. He was enamored of Minna Timmins, daughter of George Henry Timmins, a less prominent member of the local aristocracy, and had heard, most incorrectly, that Lowell was also courting the lady. Spying his supposed competitor approaching the house, Chapman burst out and started belaboring Percy with his walking stick. In due course, Lowell was rescued by a number of other guests, his injuries cared for, and Chapman subdued.[1]

Several months after Percival Lowell returned from his final Far

---

[1]Chapman was clearly somewhat deranged. He returned to his apartment in Cambridge and held his left hand inside the stove until it was sufficiently charred that it had to be amputated. In due course Minna married him anyway.

Wrexie Louise Leonard as she appeared in 1895 when first employed by Percival Lowell

Eastern travels and settled again in the Boston area in the fall of 1893, he engaged the services of Wrexie Louise Leonard as secretary. She was some thirteen years younger than he, a native of Troy, Pennsylvania, and had moved to Boston with her older (by five years) sister, Laura, where she lived at 131 Newbury Street. On 10 December, 1895 she accompanied Lowell's party to Africa on an astronomical site survey, for there was still considerable doubt that Flagstaff would be satisfactory as a permanent location.

Thereafter, Wrexie was more or less continually in the company of her employer for the rest of his life, traveling with him frequently across the United States and to Europe. The efficiency of her secretarial work is attested by the volume of Percival's literary output during these years. Historian David Strauss has compiled a list of Lowell's literary output. In addition to that noted in the appendices, he produced over 250 articles, the great majority on astronomic topics (dozens of them in German or French) and countless prepared speeches—all with Wrexie's help. Though the intimacy of their relationship was never the subject of public comment, some telltale evidences exist that it was both deep and respectful. In his absence from Mars Hill, she was obviously in charge.

Her letter to Percival of Sunday evening, 23 June, 1907, is eloquent:

Although you have not reached Boston as yet—I must be writing or you will get ahead of me on your return journey. Am fresh in from gardening—8 p.m.

Wrexie Leonard enjoying the snow on Mars Hill in 1905

Lowell Observatory staff and servants assembled in front of the Baronial Mansion in 1905. From the left are: Mrs. D.N. Petty, D.J. Termansen, Percival Lowell, Wrexie Leonard, Miss Bassett, C.O. Lampland, Mr. and Mrs. V.M. Slipher (in doorway), Mr. and Mrs. Harry Hussey, Gormley (driver), and John Duncan.

and as I turned to lock the gates who should I see but our tame friend: Jack Rabbit, Esqr. He came up from our lane and crossed the road by the tall pine trees just west of the garden and almost stopped and looked at me. He was going very slowly and finally vanished from sight in the pineys S.W. of the garden. I wanted you then, to see him too!

The gourds and pumpkins are doing as well as can be expected in the absence of their doctor. I gave them a drink tonight and I think they liked it. A lot of new ones are pricking through. Last night I covered them all for fear of frost as I don't want anything to happen to them while you are away.

The Concord gourds in the pot have flowered and their blossom looks just like a primrose—lovely! Two morning glorys were out this morning and the sweet pea under the servant's window has put forth a beautiful pink flower.

Mr. and Mrs. [Judge Edward] Doe lunched here today and how we did miss your "vacant face." It was very nice except that they staid [sic] too long—not leaving till nearly 6. Mrs. Doe sent her love to you.

Wrexie's bedroom in Lowell's Baronial Mansion, 1906

Yesterday, i.e. Friday night, the seeing was 4 and the same last night, so Mr. Lampland informs me. He said that Mr. Duncan[2] made a very good drawing of Mars. I am going to ask him to leave a copy of it for you. Mrs. Duncan came up last night and was taken ill and had to be carried down in a carriage after midnight.

Mr. Williams[3] has found an asteroid on one of those star cluster plates of June 12 (not a refractor plate) and as he cannot trace it in the book—he thinks it is new. He showed it to me and it did look like one—though a very short tail.

Professor Taylor[4] has got up a cañon excursion for tomorrow and Mr.

[2]John Charles Duncan (1882-1967) was the first Lawrence fellow at Lowell Observatory and authored the basic college textbook, *Astronomy*.
[3]Kenneth Powers Williams (1887-1958) was at Lowell Observatory as the second Lawrence fellow in 1907. He returned to Indiana University, taught mathematics and celestial mechanics and later became an authority on Civil War history - *Mr. Lincoln Finds a General*.
[4]Almon Nicholas Taylor (1855-1937) trained as a lawyer at the University of Michigan but switched careers to education and moved to Flagstaff for health reasons. He served as head of Northern Arizona Normal School (now NAU) for ten years after its founding in 1899.

Williams has decided to go—after consulting me first. And I am letting Ernest and Mary [household servants] go too. They will enjoy it and I doubt if they would have another opportunity when you are here . . .

The papering is done in the entry and tomorrow the stairing is to be done and the shelves put in. Mr. Lampland finished my Mars photographs last Saturday P.M. So tomorrow I hope to get on with that.

I am sending you all the mail of account up to date. Two enclosures with this and a fat one from Mr. Notman.[5] *Your telegram and post card from Albuquerque were a joy to get. So glad you discovered a drawing-room . . . and that you found congenial friends on board.*

The house seems very big and quiet without you and it is consoling to think that you are going to hurry back with your LLD![6] *I do hope that you will be here by next Sunday. W. L. L.*

Percival Lowell was in good company at the Eighty-sixth Annual Commencement Exercises at Amherst College. Also receiving honorary degrees were: Edward Clark Potter, a dropout from the class of 1882, who had made good as a sculptor in neighboring Enfield and finally got his Master's Degree; William Eleroy Curtis, an influential journalist resident in Washington, who received a doctorate in letters; William Greenough Thayer, who had taught at the famous Groton Academy and

[5]Frederick W. Notman (d. 1948) was a factotum in Percival's office at 53 State Street in Boston.
[6]The Amherst commencement was on Wednesday, 26 June. Lowell received his honorary degree from Dr. George Harris (1844-1922), a distinguished Congregational clergyman, himself the recipient of a half dozen such honors, who served as president of Amherst College for thirteen years after 1899.

This 1916 pumpkin from Lowell's garden measured 41 inches around the waistline.

was now headmaster of St. Mark's School, became a doctor of divinity; Charles Smith Mills, another of the Class of 1882, now a distinguished Congregational preacher from St. Louis, was also given his D.D.; Herbert Levi Osgood, Class of 1877, and then professor of history at Columbia University, received a doctorate of laws; and Anson Daniel Morse, who had been professor of history at Amherst for thirty years, also received his LL.D.

Press reports[7] noted that Lowell's citation was as "astronomer and author; distinguished for contributions to astronomical science, which have honorable recognition by numerous foreign and American societies . . . Author of valuable books on Korea and Japan . . . Liberal patron of scientific research." Undoubtedly, though unstated, a further consideration in the decision to honor Lowell was his generosity to professional astronomers associated with Amherst College, such as Professor David Todd, chairman of the department.

Three years earlier, on 1 June, 1904 Wrexie had been admitted to membership in the Societé Astronomique de France. As member #3771, this was hardly an indication of anything notable, but her sponsors were of the first rank: Percival Lowell and the grand old man of French astronomy, Pierre Jules-César Janssen (1824-1907). Lowell was in Paris at this time to accept the Societé's Janssen Medal for his work on Mars, and the

[7]Carried in the Holyoke *Transcript* of 26 June, 1907 and the Springfield *Republican* of the next morning.

Lowell preparing to play Santa Claus for Flagstaff children in 1908

fact that one of the most prestigious names in French science was attached to Wrexie's application was a signal honor. Janssen, crippled from a childhood accident, was the director of the French National Observatory at Meudon, southwest of Paris. In 1893 he arranged to establish a multi-discipline observatory almost on the crest of Mont Blanc, to which he was carried twice by teams of porters. Janssen was also the world's most highly regarded authority on eclipses.

The "Baronial Mansion," in which both Percival and Wrexie lived when on Mars Hill, was a rambling, somewhat jerry-built, split-level structure that had been expanded by a series of interconnected afterthoughts from the four-room "cottage" of 1894. The original structure had been placed one hundred feet north of the "Clark" dome, in order that no currents of escaping warm air might interfere with the seeing. In 1902 it had been remodeled with an additional six rooms and by the time of Percival's death it had eighteen.

At its apogee, the servants' quarters were at the extreme northwesterly end, adjacent to the garage, then three master bedrooms (one of which pertained to Wrexie and another to Percival) surrounded a reëntrant porch opening to the northeast. Bathrooms, kitchen, dining room and other bedrooms opened onto the south and ultimately crowded even onto the roof of the library below to the east.

Percival Lowell's concern for local wildlife led to the staff making a suitable decoration on the cake prepared for his 53rd birthday party.

The original cottage had contained a study, two bedrooms, a workroom and a darkroom. Underneath, on the northerly down slope of the hill, were a bathroom, another darkroom and a wood storage area. Entrance to this area was via a trap door in the floor of the northeast corner of the workroom, just in front of the easterly bedroom door. One night Lampland forgot to close the trap door after leaving the basement and when V. M. Slipher got up the next morning, he came out of the east bedroom door and was badly bruised by falling into the darkroom. Later, when the house was expanded, the workroom became the dining room and parlor. The lower darkroom became the wine cellar.

Godfrey Sykes built various extensions to this cottage over several years. The first additions were made to the west and the bathroom next to the study was unique in that it contained a sunken bathtub, long enough that a tall person could lie at full length in it. However, it was set so that only the curved rim of the tub was above the floor and one could easily tumble into it. That expansion project also included the later much talked-about "secret passageway" which went below the main floor from just outside the doorways to this bathroom and the study, in a northwesterly direction, to emerge on the porch just outside the doorways to the three master bedrooms. It was about three feet wide and high enough so that one could walk upright, with stairs at each end to reach the floor

Winter view of the front facade of the Baronial Mansion in 1911

Percival Lowell trudging through the snow on Mars Hill in 1915

Wrexie Leonard, zoologist Edward S. Morse, and Percival Lowell in the office above the original library of Mars Hill in 1906

level. En route to the porch, the passageway crossed under portions of the kitchen, butler's pantry, den and Percival's bedroom; it provided a means of access to the bathroom without the necessity of passing through other rooms of the house.

The evidence of Wrexie's devotion to Percival is hard to escape and is most clearly visible in the small volume she published several years after his death; *Percival Lowell—An Afterglow*. There is no mistaking the point touched on sixty years later by Jan Michael Hollis in his monograph entitled *Wrexie*.[8] Citing her dedicatory poem:

> *Preambient Light*
> *Waning, lingers long*
> *Ere lost within.*
> *Just, kind, masterful:*
> *Life's sweet constant,*
> *Farewell.*

[8]See Griffith *Observer*; Vol. 56-1; January, 1992.

Percival Lowell studying negatives on the north porch of the Baronial Mansion in 1914

He continues:

One may easily guess that Preambient Light is Percival Lowell, but deciphering the next line is more difficult. Diminishing light is inferred, maybe even intended for the casual reader, but Wrexie Louise Leonard (as Waning, Lingers Long) is meant. Wrexie had carved two sets of full initials upon this tree! ... The Farewell to Percival and the all-too-swift passing of his light from her life, is poignant.

But Hollis glimpsed only a portion of the eloquence hidden in Wrexie's verse. Lines 3 and 4 also contain initials, those of two other long-time and loyal employees of "Preambient Light," Elizabeth Langdon Williams and John Kenneth Macdonald, the "computers" who also wit-

Percival Lowell and Professor E.S. Morse on Bright Angel trail in the Grand Canyon

nessed portions of his will and did the bulk of the calculations that ultimately led to the discovery of Pluto.[9]

Line 5, however, has a complete reverse twist—the initials are of the final major player to affect the life of Percival Lowell, and the one who almost succeeded in completely negating his life's work. Every aspect of line 5 can be safely read in the reverse of its apparent meaning—as can the initials themselves.

[9] Both stayed on after Lowell's death, Miss Williams (1882-1981) until 1922 when she married another staff member, George Hall Hamilton (1884-1935), and then departed Mars Hill when he took a job at Harvard's observatory in Jamaica. Williams was the major exception - until recent years - to the all-male appearance of scientific professionals on Mars Hill and elsewhere in the world of astronomy. Macdonald was her principal associate and also served as Lowell's bookkeeper.

Percival Lowell at breakfast on the porch of the Baronial Mansion in 1913

Sadly for almost all concerned, in 1908 Percival abruptly decided to marry his neighbor from around the corner in Boston, Constance Savage Keith. This was no raving beauty or gloriously rich prize; she had been in the business of rehabilitating run-down housing in the Beacon Hill/Back Bay area of Boston. Percival had even financed one such endeavor for her in 1896, but there was no record or even rumor of his prior romantic interest. However, in a special dispatch to the *Boston Herald*, datelined New York; June 10; one reads:

> *Percival Lowell, astronomer and orientalist, married Miss Constance Savage Keith, daughter of Beyrer Richmond Keith of Boston, this afternoon in St. Bartholomew's Church, Madison avenue and Forty-fourth street. Only a few intimate friends were present at the ceremony, which was performed by the Rev. Leighton Parks, rector of St. Bartholomew's.[10]*
>
> *The bride, who was unattended, wore a frock of white Liberty satin, trimmed with Brussels point and a white Tulle veil. She carried a shower bouquet of sweet peas . . .*

[10]Interestingly, Pauline C. Hamme, the parish secretary in 1991, could find no record of the ceremony in the church files.

Percival Lowell feeds a visiting dog at his front door in 1912 . . .

. . . and exhibits a gopher snake that inhabited the area between the Mansion and Clark dome the same year.

The age of the bride, according to the marriage license which was taken out at the City Hall this morning, is 44 years and that of the bridegroom is 53.
BUT FEW INTIMATES IN BOSTON KNEW OF INTENTION TO MARRY . . .

The article went on concerning the unexpected nature of the marriage. However, word was slower in reaching Flagstaff, where the *Coconino Sun* reported the event under date of 19 June, giving much of the same information but concluding on a more astronomical tone:

. . . They left immediately on a honeymoon tour for Canada and the northwest and are expected to arrive here some time next September. Information concerning the wedding is meager, and it can only be conjectured that the study of Mars will be superseded by a study [of] Venus.[11]

At this distance in time, one can only speculate as to why Percival, one of the more eligible bachelors in the nation, chose to marry a woman who was by no stretch of the term a beauty and, at age forty-four, could hardly be counted on to give him an heir. Matrimonially speaking, she had been "on the shelf" for years and there exists no record of Percival's thinking on this matter. Was there some overriding element of family pressure? His mother had died in 1895, his father in 1901, and his sisters do not appear to have been able to exert serious social pressure. Katherine, three years younger, was by now removed to Philadelphia; Elizabeth, seven years younger, was apparently uninvolved; and Amy, nineteen years younger, was almost of a different generation.[12] Did Lawrence, by now president of Harvard and a paragon of respectability, persuade him to take a wife?

No matter which way the question was sliced, marriage to Constance Keith was hardly a social upgrade for a Lowell, while Wrexie so clearly was of great help in his writings and had afforded him much personal support. The eloquence she helped him produce was magnificent (See Appendices), of which the following passage from *Mars and its Canals*— graven into the granite of his mausoleum—is typical:

---

[11] Announcements were mailed from Boston two weeks after the event:
"Mr. and Mrs. Daniel K. Kane have the honour to announce the marriage of their [sic] sister Constance Savage Keith to Mr. Percival Lowell on Wednesday, the tenth of June, nineteen hundred and eight, at St. Bartholomew's Church, New York City."
[12] In later years, Elizabeth was quoted by her daughter-in-law, Caroline (Mrs. Roger L. Putnam), as having observed, with reference to Percival; ". . . my brother was a brilliant man, but not a good one."

Astronomy now demands bodily abstraction of its devotee . . . To see into the beyond requires purity . . . and the securing it makes him perforce a hermit from his kind. He must abandon cities and forego plains . . . only in places raised above and aloof from men can he profitably persue [sic] his search . . . He must learn to wait upon his opportunities and then no less to wait for mankind's acceptance of his results . . . For in common with most explorers he will encounter on his return that final penalty of penetration the certainty at first of being disbelieved.

The pictures extant show Wrexie as an attractive woman—but she was Lowell's secretary and proper Bostonians just did not marry their secretaries.

Constance's own argumentative and contentious nature may have been a contributing factor in cementing the relationship. Though there is little direct evidence that she might have blackmailed Percival into matrimony, her contentiousness was to become the cause of considerable embarrassment to her husband who found himself irritated by lawsuits regarding her ongoing business activities in Boston. Certainly her subsequent record of legal maneuvers to retain control of his estate showed a seldom equalled record of sustained litigation. In the words of a later generation, she was "one tough cookie."

Seen in hindsight and with the knowledge of events occurring after Percival's death, his marriage was a disaster of the first magnitude for the pursuit of astronomy ". . . at my observatory in Flagstaff . . ." But Wrexie continued on as Percival's secretary, obviously a cause of some irritation and insecurity to his wife for, within hours of his death, her employment at Percival's office was abruptly terminated. She was paid $250 and ordered off the premises.

Wrexie's final years were lost to Percival's successors on Mars Hill. Percival's official biographer, his younger brother Lawrence, left her completely out of the record, dwelling on several other employees, like Williams and Macdonald, who had a long association with the astronomer, but not the one who had worked for him the longest and the most devotedly. However, it is obvious that Percival and Wrexie maintained a warm relationship during their years together on planet Earth, as is made clear in her small monograph, but no copies remain of his half of any correspondence and few of her communications survive in the files of Percival's observatory.

The bad blood between Constance and Wrexie was of long standing. Exactly four years prior to her husband's death, Constance had written to V M:

. . . I am sending you the key to my room too. I did not before because I thought Miss Leonard was coming and I knew the key would be asked for. I thought it best for you to be able to say you did not have it. If she comes now before I do, please be kind enough to say that I did not want my room opened until I came - or died, and in that case you or Mrs. Lampland are the only ones I would wish to attend to what things I have left there . . .

Wrexie returned to the East following her dismissal from Mars Hill and seemed, according to her grandnieces, to be in comfortable financial condition; but all this changed after the market crash of 1929. She lived for a while with a niece in New York and in her final years resided in Roxbury's Trinity Church Home for the Aged. In 1937, shortly before her death, she was moved to the Medfield (Massachusetts) State Hospital. Significantly and poignantly, during these final and most humble years of her life, Abbott Lawrence Lowell, who managed to totally ignore Wrexie Louise Leonard in the formal biography of his older brother, contributed regularly but quietly towards her support.

The lady that became Mrs. Lowell, however, was cut from far different cloth. Upon the death of her husband, by stroke on 12 November, 1916, she sought to assume command of his observatory and did so with considerable success. The designated trustee under the ruling codicil of Percival's will was his eldest nephew, George Putnam, but Constance was able to produce enough witnesses to an unwritten change of heart on the part of her late husband that George rapidly opted to remove himself from the fray.

The will, executed on 21 February, 1913, granted broad powers to the executor and successor trustees, but was otherwise relatively simple and reads—in pertinent part—as follows:

First, I appoint my wife, Constance S. Lowell, and my brother-in-law, William Lowell Putnam, executors . . .

Third, I give to said Constance S. Lowell the sum of One Hundred and Fifty Thousand (150,000) Dollars and all my personal and household effects and my automobile.

Fourth, all the rest and residue of my property I give to my brother-in-law, William Lowell Putnam, to be held subject to the provisions hereinafter made for my wife, in trust for the Lowell Observatory. The property shall be invested. Ten (10) per cent of the net income shall be added yearly to the principal, and the balance of the net income shall be used for carrying on the study of Astronomy, and especially the study of our Solar System and its evolution, at my Observatory, at Flagstaff, Arizona, and at such other places as may from time to time be convenient . . .

The trustee shall also have full control of the management of the trust and shall have power to appoint, select and discharge the director and all other employees and agents of the Observatory, but I hope that Dr. V. M. Slipher will be appointed the first director after my decease . . .

Each trustee shall appoint his own successor within a week, or as soon as possible after his accession to his office, in order that no failure of a regular nomination may take place . . .

Fifth, The Lowell Observatory shall at no time be merged or joined with any other institution.

Other provisions included a fallback mechanism for selecting a trustee in the event of a failure in the stated process, further provisions for the financial support of Constance, including an annuity of $60,000 per year and the right to lifetime occupancy of his house at 11 West Cedar Street (and around the corner to include 102-104 Mount Vernon Street) in Boston. In a series of codicils over the next three years, certain changes were made.

Only a week after its execution, as is common in many wills, the lawyer who drew up the initial document was replaced. Putnam's successor as executor and trustee was Percival's old college friend and third cousin, Harcourt Amory, Jr. Parenthetical to the future litigation, on the same date Constance was given an increase in the outright grant from $150,000 to $175,000. Of greatest future significance, though, was the fact that Constance's name was not otherwise mentioned in this alteration—she presumably stayed on as executrix, a point she was to press with vigor and success.

The following 22 November, Percival's nephew, George Putnam, eldest child of his sister, Elizabeth, replaced Amory as trustee, though obviously not as co-executor, an addendum endorsed by Constance and witnessed, among others, by W. Louise Leonard. And a year later, on 15 December, 1915, Constance was given the right to lifetime occupancy of the "Baronial Mansion" on Mars Hill, that had grown by an almost endless series of additions from that first little cottage to house the working astronomers in the summer of 1894.[13]

---

[13]Constance did not like the name "Baronial Mansion," preferring "Hollyhock House." But the former name stuck until the rambling structure was removed as a fire hazard in 1959.

Tsunejiro Miyaoka, a respected Tokyo attorney who had been Lowell's secretary at the age of 17 in 1883, visited his grave on Mars Hill in 1917. He was driven from the Flagstaff railroad station in Lowell's Stevens-Duryea.

The Explorers of Mars Hill / 88

The death of Percival Lowell set in motion a wholly different scenario at his observatory. Meanwhile, the event was noted in newspapers around the world and memorialized in several astronomical and other journals. Among all these tributes few came from associates more aware of his presence than the Appalachian Mountain Club, of which he was the president at the time of his death.[14] The Club's Council met nine days after Lowell's death, with Fred Harmon Tucker presiding, and adopted a resolution prepared by William Witherell:

> WHEREAS the late Percival Lowell was the President of the Appalachian Mountain Club, and
> WHEREAS the late Percival Lowell was an active member and upright gentleman beloved and respected by all, and
> WHEREAS the late Percival Lowell by his deep study and unsurpassed work in astronomy attained a position among his fellow men whereby he was regarded by the Appalachian Mountain Club as one of its most distinguished members,
> BE IT RESOLVED: that in the untimely death of Percival Lowell the Appalachian Mountain Club as a club and individually has suffered the irreparable loss of a leader and true friend, who by his works and deeds added to that which is best and worth while in the lives of his fellow members.

More emotional in its message, however, was the composition of J. Corson Miller that was reprinted in the Chicago *Tribune* under date of 6 February, 1917:[15]

THE DEPARTED ASTRONOMER

Across the gentle night stars bud and bloom,
Tolling the ebb and flow of cycling time:
Spun out from the Creator's mighty loom,
They sing forevermore the Ancient Rhyme.
Purple and gold and bluish-white they gleam,
Across these crags and canyons, thunder-sown.
The garden-paths of Pollux lie a-dream,
While Death - the Sentinel - keeps watch alone.

[14]In its earlier years the Appalachian Mountain Club was a very scientifically oriented organization, dominated by professors at MIT, Harvard and other like institutions in the Boston area. Its first president was the elder Pickering brother and the younger also came to hold that office. Lowell had presided at only one Club meeting during his term.
[15]The place and date of original publication are unknown, but the poem appears in Miller's 1921 volume, "Veils of Samite."

Lo! He is gone - the Searcher of the Skies.
No more the mountain breezes stir his hair,
The while he marks, with genius-flaming eyes
The hills on Mars, or some young comet's lair.
Great curving streams of suns and wreaths of stars
That swung before him in that fiery sea,
Now play his funeral march on viewless bars -
Aerial Ocean's proudest pageantry.

Yea, he is gone! Yet somewhere, with the sun
That scatters far the laggard mists of morn,
His spirit soars like Rigel - Silver One,
Whose colors oft the blue eastern Night adorn.
Not by lone trapper's trails, nor on the sea,
Nor in the woods, when Evening's lamp burns dim,
Shall he be met, but 'mid the galaxy,
Of Suns, and Moons, and Stars, look ye for him!

Stimulated by Constance's prompt assumption of control and a rapid-fire series of telegrams to her in-laws back in Boston, on 13 November, 1916, the very day that news of Percival's death was received there, Harcourt Amory sent a letter;

To The Judge of the Probate Court,
Flagstaff,
Arizona.
I hereby decline to act as either Executor or Trustee under the will of Percival Lowell, late of Flagstaff.

This curt resignation was signed in the presence of Edward S. Flynn, one of Constance's Boston agents, and Harvey Hollister Bundy, attorney and since early 1915 son-in-law to William Lowell Putnam, II. Short and sweet to her eyes, this letter was important for the widow's ten-year effort to retain the entire estate in her control.

Relying on Constance's statement that her husband had planned a codicil removing George Putnam as Trustee-designate in favor of Guy Lowell, Percival's third cousin, George followed up with a letter that had obviously been prepared for his signature:

> *To the Judge of the Superior Court of Arizona in and for the*
> *County of Coconino*
> *In the Matter of the Estate of Percival Lowell,*
> *late of Flagstaff, Arizona.*
> *I George Putnam, of Manchester, Massachusetts, nominated as trustee under*
> *will of said Percival Lowell, hereby decline to act as such trustee and request that*
> *such declination be accepted.*
> *Dated December 19th 1916.   Geo. Putnam [s]*

This action left the office open with no successor on the horizon. There is no record of George's reason for declining but he had ample reason; he was newly married, struggling to start his ultimately noteworthy career in the brokerage business, and a Reserve Officer in the U.S. Army with America's entry into World War I clearly in the offing.

Constance continued her campaign for someone she felt would be scientifically unimpeachable and if not complacent, at least congenial. But, in keeping with the fallback provision of Percival's will, the task of finding a replacement now devolved upon ". . . the oldest male descendant of my father, Augustus Lowell, then living, who shall appoint himself, or any other qualified person."

Meanwhile, in his obituary report to their Harvard classmates, Lowell's friend, George Russell Agassiz, had concluded with undue optimism: "His entire fortune, with a certain life interest for his wife, was left to maintain the Lowell Observatory as a separate institution. It is thought that its income will eventually be at least twice that of Harvard Observatory."

Abbott Lawrence Lowell, younger by one year than Percival, was thus the designated decision-maker. However, as sole trustee of the Lowell Institute and President of Harvard University, he had his plate full. Constance had already made it plain that despite anything Percival had left in writing she felt she was the only valid successor to her husband. It was a difficult spot, but George's withdrawal had made his part necessary and Lawrence had to face the messy issue. He opted for peace with the widow by naming her preference and filed the appropriate papers with the American Academy of Arts and Sciences on 21 December, 1916, which were then transferred to the Court in Flagstaff and duly approved on 10 January, 1917.

Thus the stage was set for a ten-year legal epic.

6

Lowell vs Lowell
1916 – 1927

E NTER: Guy Lowell, Percival's third cousin, descended from the second marriage (Cabot-Jackson line) of the Old Judge.

Born in 1870, the son of Edward Jackson Lowell, II, and Mary Goodrich Lowell, he was educated in Dresden, then a famous center of artistic learning, and Paris. He entered Harvard with the Class of 1892 and attended MIT for two years after graduation. There, he studied under the distinguished Franco-American landscape architect, Constant Désiré Despradelle, only eight years his senior. Guy went on to the École des Beaux Arts at Paris for further study, from which he graduated in 1899, having married Henrietta Sargent the year before.

Finally, in 1900, he hung out his shingle in Boston and was immediately in demand. The fruits of his subsequent brilliant work can be found today in countless private homes and estates of Boston's North Shore and in such public places as Andover Academy, the diminutive but striking Public Library of Boscawen, New Hampshire, the Cumberland County Court House at Portland, Maine, the massively colonnaded Boston Museum of Fine Arts and the imposing granite of the New Hampshire Historical Society in Concord.

Guy put in two years of distinguished service in the final years of World War I, holding the rank of major and directing Red Cross ambulance workers and hospital attendants. The quality of his service was attested by the decorations he received subsequently from both the

French and Italian governments. A great part of this service was among the Alpini in the mountains of northeastern Italy where some of the most interesting phases of World War I were fought. But his absence on national service during these years also played into the self-serving hands of Percival's widow.

Guy Lowell designed and laid out the Quadrangle of Radliffe College; he worked for Harvard, too, designing several buildings, of which the "New" Lecture Hall probably became best known to most students. This building was funded by a gift from Lawrence Lowell and served from the day of its completion as home to his (and his successors') lectures in government. However, Guy's most famous landmark is unquestionably the New York County Court House in lower Manhattan. Its octagonal shape alone is unique in such structures; but he won the job in competition with seventy entries, the top ten of which were then reentered into a second round of competition with the ten most notable architects of the country. Started in 1913, it was completed fourteen years later at a cost of $20,000,000. Its dedication, early in 1927, came one week after the death of its designer.

Distinguished as he deservedly was among architects, in Flagstaff Guy Lowell found himself in a vastly different and less rewarding arena, from which he and his successor were fully extricated only after another thirty-eight years. From the outset of his responsibilities in Arizona, Guy found his conscience and his duty at loggerheads with his cousin's widow's determined personality and grasping nature.

With her late husband's friend, Judge Edward Doe,[1] as attorney, Constance set out within a week of Percival's death to circumvent her husband's will. Rumors soon reached the staff that she was intent on breaking it completely, so that she would become the sole heir and beneficiary.[2] Upon motion of Judge Doe, however, the will was filed and the court endorsed Constance in her status as executrix. Doe promptly went for the jugular and immediately entered the argument that the law of Arizona forbade perpetuities. But that first effort to obtain the whole estate for herself was rapidly settled in favor of the defendant trustee.

One of attorney Doe's next pleadings to the Court was to increase the widow's monthly stipend from her opening request of $300,[3] contending the sum

---

[1] Doe was a sometime associate justice of the Arizona Supreme Court.
[2] Carl Lampland's diary notes this under date of 11 January, 1917.
[3] Seventy-five years later, this would have represented a purchasing power on the order of $6000 per month.

*Charcoal study of Guy Lowell, the first trustee of Lowell Observatory
by John Singer Sargent*

. . . is not sufficient to support and maintain said widow in the manner and style she has heretofore lived, nor is it sufficient to support and maintain her in a proper manner while she is conducting the settlement of the estate, and asks that an allowance of $1,500 per month be made . . .

On 30 March, 1917, two months after the filing of the will, Judge F. W. Perkins raised her take to $1,200 per month.[4]

Having lost the opening skirmish on one front, however, Constance was not out of the war and now switched her strategy so as to perpetuate her position as executrix. For the next four years she filed none of the required probate reports, but repeatedly told the court that she was diligently trying to wrap up the affairs of her late husband. The court was tolerant, as in the following order, typical of several that followed over the next few years:

IN THE SUPERIOR COURT OF THE STATE OF ARIZONA
IN AND FOR THE COUNTY OF COCONINO

IN THE MATTER OF THE ESTATE OF
Percival Lowell,
Deceased.

Constance S. Lowell, Executrix herein, having this day filed a statement, supported by affidavit, showing why she is unable to make an annual settlement and praying that she be excused from making same until she is able to procure the necessary information and vouchers, as set forth in her said petition; and the Court being satisfied with the explanation made;

IT IS ORDERED, That the said executrix be excused from at this time filing her annual report.

Done in open Court this 11th day of December, A. D. 1917.

F. W. Perkins
Judge.

Meanwhile, without rendering an accounting to anyone, Constance continued to happily pocket her very substantial widow's allowance as well as half the estate's income.[5] Besides her own commutation to and

---

[4] She also had two rent free houses, her husband's automobiles and control of his estate. The staff astronomers' annual salaries were on the order of two months' worth of her cash allowance.

[5] She even charged the staff astronomers for copies of Lowell books they had helped write. On 18 March, 1920, V M was billed $17.40; Lampland, $21.60 and E C, $20.30 - all of which were paid promptly.

Judge Doe on his way to court in Prescott, 1919

from Boston, she also paid for frequent cross-country trips for her niece and other relatives out of Observatory funds. All this understandably caused Guy Lowell enormous frustration in trying to exercise his responsibilities as Trustee. He soon changed from his initial stance as a congenial and sympathetic relative to that of an equally determined and strong-willed adversary.

The contentious proceedings were not quite as lengthy or as debilitating to the estate as in the Dickens' classic of "Jarndyce vs Jarndyce,"[6] but their existence and the continued bickering did much to hamper the effective utilization of the considerably smaller resources then available "... for carrying on the study of astronomy ... at my Observatory, at Flagstaff ..."

The staff, meanwhile, caught between a largely resident executrix whom they knew, and a largely nonresident trustee, whom they did not know, played both sides of the street. When importuned by Constance, they signed statements she drafted, but otherwise tried to stay out of it

[6]See *Bleak House*, 1853.

The Explorers of Mars Hill / 96

Coconino County Courthouse in Flagstaff, 1916

and do their scientific thing. Miraculously, despite frequent lapses in pay, they performed quite credibly during this exasperating and drawn-out interregnum.

To the staff, Mrs. Lowell exhibited many idiosyncracies. After her husband's death she only wore black clothes, a black hat and black veil or shawl for the remaining thirty-eight years of her life. She also complained of failing eyesight and when anyone was in her presence she kept her eyes downcast so that the pupils were largely covered. However, on one occasion when Bernice Giclas brought her a cake, she commented on it while some ten yards distant. The staff always tidied up around Percival's mausoleum when she was expected, for she invariably asked for the key and would spend an hour or more there, several times during each visit. She would emerge distraught after these sessions in the mausoleum and mumble about "being sustained only by the spirit." During her lifetime, she carefully preserved an inscription on the wall of his bedroom, written in chalk, that stated, "Percival Lowell's earthly existence terminated in this chamber upon the green couch."

During these visits, the staff never knew which bedroom the widow would occupy, and there was strong evidence that she never spent an entire night in any one of them, the blankets and sheets being moved from one room to another during the night. Though Constance had the right of lifetime occupancy for this building, she spent nothing on its upkeep (what few repairs were made came out of the other half of Percival's estate) and the place gradually deteriorated.

Over the decade following the death of its founder, the divisiveness affecting the little community atop Mars Hill, west of Flagstaff, was played out in periodic acts heard mostly in the Coconino County Courthouse below. Members of the scientific staff found themselves involved as witnesses to Percival's original intent and were also subject to periodic interrogation by Mrs. Lowell while at work.

Despite the fact that her original counsel, Edward Doe, had filed the will on her behalf, and that she had sworn to be its true and faithful executrix, on 7 January, 1921, four years after rumors first reached the staff of Constance's desire to break the will, exactly such a plea was again made on her behalf in court. In 1919 Doe had fortuitously died, thus giving her new attorneys an excuse to ask for another lease of time in which she could maintain control of the estate's dwindling assets before filing this final effort at acquiring the whole of it for herself.

In the midst of the next legal delay, under date of 29 September, 1921, and after failing to get any takers in her suggestion that everyone take a three-month unpaid vacation, Constance drew up a form by means of which she notified the staff of salary reductions:

Mrs. Lowell begs to inform you, with much regret, that the funds of the Estate of Percival Lowell have diminished to such a degree, that said funds will not permit of the distribution of the present salaries in the future. You are now offered $____ per month. This does not mean, naturally, that you are compelled to accept it; no ill-feeling will be harbored if you do not, neither will any be generated by your frankness in admitting that such a salary would be out of your consideration

Mrs. Lowell requests that you will give her an answer as to what your decision is, within the next few days; she also wishes to state that she expects, if you remain, that you will do so in co-operation with existing circumstances, not as a personal favor to her nor as one to the Observatory; for neither would be acceptable, unless at the same time you felt that you were availing yourself of an opportunity equivalent to that which you were getting or might get elsewhere or otherwise.

Mrs. Lowell, at this time wishes to express her heartfelt appreciation for the faithful work you have rendered the Observatory in the past and for that which you may have occasion and opportunity to do for it in the future.

Meanwhile, in an exclusive dispatch from Flagstaff to the *Los Angeles Times*, only two months earlier, under a headline "MAUSOLEUM FOR ASTRONOMER" the public was informed,

A magnificent granite mausoleum, to cost about $40,000, is being erected on Observatory Hill, west of Flagstaff, to contain the remains of Dr. Percival Lowell, founder of the astronomical observatory to which his name is attached. The tomb will be between the observatory and the Lowell residence. It will be circular, with a dome. Three carloads of granite blocks, each of 1000 pounds weight, have already been received here for the structure. In the Lowell will ample provision was made for carrying on the observatory work, this duty being taken up by the scientist's widow.

In the January 1921 plea Constance's attorneys used the unique argument that while the Court had in 1917 affirmed her status as executrix, on the petition of Judge Doe, and she had been prompt to exercise her privileges under that status and most unwilling to relinquish them, in fact the court had decided the matter of the ultimate purpose of the estate quite improperly. Counsel for the Trustee had to file a fifty-page brief with the court to turn back that motion, but Constance won further delay since the Court took until 8 July, 1924, to hand down its decision.[7] During this period, Messrs Crable & Cornick, attorneys for the Trustee, continued their research into the pertinent statutes, including a month-long stay to pore over documents in the Cleveland, Ohio, Municipal Library.

One of the more inventive contentions used by Constance in this "get rich the easy way" effort was a contention filed in 1924 to the effect that she had made an error in her first probate report. When her initial case against perpetuities seemed likely to hold up, less than 20 percent of the total assets listed in Percival's estate were under the heading "Community Property." But when that pleading was dismissed it turned out that over 80 percent of the assets had been "Community Property"—a massive alteration and "oversight" that the court also found difficult to swallow.[8]

In the intervening years, Constance lived frequently on Mars Hill,

[7]William Lowell Putnam, II, who had drawn up the much contested will, patterned after that of John Lowell, Jr., had died one month earlier while visiting his other brother-in-law, A. Lawrence Lowell, on Cape Cod.

[8]With the benefit of hindsight it seems likely that much of the legal inventiveness displayed in this effort stemmed from the clever and fertile mind of her nephew, R. Keith Kane.

though feeling free to commute to Boston as convenient, often accompanied by one or more of her sister's children, who were frequent visitors to the Observatory.

In the end, the Coconino County court found for Guy Lowell, in his official capacity, and ordered that the estate be distributed in accordance with the will. But Constance immediately appealed, though her attorneys prepared a record so defective and erroneous that they were compelled to submit two amendments. This time, Messrs Crable & Cornick submitted one hundred ninety pages of reply, including citations of hundreds of prior cases supporting the Trustee's position.

As Guy's legal noose slowly tightened, Constance kept seeking to evade the day of reckoning. Even from far away in Massachusetts she carried on her campaign, swearing an interesting and self-serving affidavit which her attorneys filed in Flagstaff on 24 August, 1924, seeking yet another delay:

> Constance S. Lowell, being first duly sworn, deposes and says that she is the Executrix of the will of Percival Lowell, deceased, and that she has received the context of the order of the Probate Court of Coconino County, Arizona, requiring her to file a supplementary and final account as such Executrix on or before September first, 1924; that she has since been diligently engaged in the preparation of the final account, but in view of the fact that it must cover a period from October 15th, 1920, the date of the filing of the last account, to the present time, it is impossible for her to complete said account by September first, but she believes she will be able to complete said account by October first, 1924. Deponent further states that the additional time required to prepare this account is partly due to the fact that errors have been made in previous accounts in reporting correctly Community Property, all of which entails a correction of heretofore reported income thereon.
>
> And further Affiant saith naught . . .

At long last, on 9 October, 1925, five months after Percival's youngest sister, Amy, had been laid in her grave, the Arizona Supreme Court handed down its decision, upholding the validity of the trust, which Constance's lawyers had twice attacked on the hoary legal maxim against "perpetuities." The Court continued to find nothing wrong with a perpetuity in the public interest.

Thus finally stimulated to performance, the contending parties arrived at an agreement on 15 December, 1925, in Boston, by which the residual issues were largely settled. The Estate, now dwindled to less than half its original size, by "wastage," legal fees, taxes and executor's fees, was turned over to the Trustee's administration, under which it began the long, slow climb back to utility. A much smaller separate trust was created from some of the funds Constance had improperly extracted from the estate, one-quarter of the income from which was for the Observatory and all the principal of which was to accrue to it upon the death of Constance. She was given "... all right, title and interest to any copyrights or scientific papers of Percival Lowell, it being understood that Mrs. Lowell will leave the said papers and copyrights to the Observatory upon her death."

The final argument concerned the appropriateness of the Trustees legal fees. Constance had shown no hesitation in making payments of $13,000 to the estate of Edward Doe for his initial filings, but she took great umbrage at the bills of Crable & Cornick for the Trustee and forced another round of expert testimony and depositions solely to determine the fairness of those fees.

In his overview of the Lowell family, Ferris Greenslet (1875-1949), among other errors of fact, observes that much of Percival's once-substantial estate had diminished "... owing to the large number of shares it contained in futuristic enterprises."[9] However, the court-appointed audit of his estate, made by disinterested parties within a few weeks of his death, is in considerable detail and reveals a very conservative emphasis in the portfolio. The major holdings were in a variety of utility bonds and then stalwarts such as A T & T. Its net value on 5 February, 1917, was $2,336,813.77. By the time Constance was through with it, almost nine years later on 30 November, 1925, an equally detailed report showed it to have a net value of $1,105,017.08.

Constance was still slow to let go. In a letter to V M of 30 March, 1926, she enclosed a check for $9500, explaining that some of it was to make up for previous pay cuts:

... After our conversation I realize more than ever the close margin under which you have been sailing. The added cost of living being responsible. In order that the denials in the future may not be as strenuous I raised your salary again January 1st, 1926, 5% making it then, a year, $4365.90 - $363.82½ per month. With the 20% raise in 1923 Mr. Lampland's now is 3960.00 per year, 330 per month.
Mr. E. C.'s $3120.00 per year - 260.00 per month.
Let Mr. Sykes pay be on a salary basis of $2640 per yr.

[9] *The Lowells and Their Seven Worlds*, page 367.

. . . feeling and realizing that even what I have done though practically a necessity, must be deemed subject to the Trustee's approval. That you all have kept pent up for so long under the hardships you have had and endured, you have my highest regards and deep appreciation. To tide over the crisis I have done all as Executrix, I feel I should. When the Trustee takes charge tell him your views and desires. I feel sure that if it is in his power to grant them you will have them allowed.

In final result, from an estate that had a princely value at the time of Percival's death, the trustee received the net sum of $1,250,000.00. This during a time of great market value increase in the American economy. Constance also pocketed a tidy balance for her final fee as Executrix—$68,592.55[10]—and maintained the right to continued occupancy of a house in Boston as well as on Mars Hill. The last paragraph of the settlement agreement reads:

Constance S. Lowell is to have the right to complete at her own expense the work she has planned to do on the Mausoleum of her husband, and the said Guy Lowell as Trustee, so far as he may lawfully do so consents that said Constance S. Lowell, if she so directs by will or otherwise, may have the right of burial in said Mausoleum.[11]

The small trust thus established was to be managed by two trustees; Guy Lowell designated the Merchants National Bank of Boston,[12] and Constance selected George Read Nutter (1863-1937), a childhood contemporary who had been president of the Boston Chamber of Commerce in 1920 and headed the Massachusetts Bar Association in 1927. A Phi Beta Kappa graduate of Harvard in 1885, he had been a law partner with the ruthlessly competent Louis Dembitz Brandeis (1856-1941) during the twenty years prior to the latter's appointment to the United States Supreme Court in 1916. Nutter was sufficiently prominent in his own right to rate a front page obituary in the *Boston Globe* at the time of his death in late February 1937.

[10] One wonders how this must have appeared to Andrew Douglass, who really labored for his $800 per year. He would have had to work for more than 85 years in order to receive just this final fee.
[11] There is also a memorial slab to Percival in the family plot in Mount Auburn Cemetery in Cambridge, just up the slope from his parents and down from his equally famous brother and sister. The slate is crowned by a massive piece of petrified wood and tells of his life, ending with the words: "REVALÁVIT MARTEM" - He revealed Mars.
[12] In time to become the major segment of the late Bank of New England.

Workmen completing the masonry on the Lowell Mausoleum, 1923

Thus things stood, more or less tranquilly as far as the pursuit of astronomy was concerned, finally allowing the Trustee to remove the mildly pejorative "Acting" under which the already distinguished V. M. Slipher had been serving as director for this troublesome decade.

To replace Nutter, Constance designated her nephew, Richmond Keith Kane (1900-1974) of New York City, as the successor trustee of her interest in this smaller trust. Kane, one of the beneficiaries of the enormous "wastage" from the estate during Constance's executorship, was also an attorney of prominence with an impressive record of government service during World War II and thereafter. A trustee of the Cathedral of St. John the Divine, he was also a Fellow of Harvard College and given an honorary doctorate of laws in 1971. Constance had picked good talent to represent her interests.

After that change, which brought no change, the very hand-to-mouth existence of Lowell Observatory continued for another seventeen years, until terminated by an event reported in the *Boston Globe* on Monday, 27 September, 1954.

MRS CONSTANCE LOWELL

Funeral services for Mrs. Constance Savage Lowell, 91, of 66 Chestnut St, Lexington, will be held at the Church of the Advent, Mt. Vernon St. today at 10. Mrs. Lowell, who died Saturday, was the widow of Dr. Percival Lowell, famous astronomer, who was the brother of poetess Amy Lowell and of Dr. A. Lawrence Lowell, president of Harvard.

Mrs. Lowell lived in Newport, R. I. until two years ago.

Her husband, who died in 1914 [sic], predicted the discovery of the planet Pluto and studied the planet Mars. He founded the Lowell Observatory in Flagstaff, Ariz., and Mrs. Lowell continued as head of the observatory after his death. She is survived by two nephews, Theodore F. Kane of Montpelier, Vt. and R. Keith Kane of New York city and a niece, Mrs. Royal Joslyn of Newport. Burial will be in Marion.

Constance's final years—like Wrexie's—were lonely. In her eighties she was described as appearing like a "benign witch," and living in "opulent squalor" in the house on Beacon Hill. Her niece, Beryl Joslyn, regularly importuned her for money and she saw little of her other relatives; those by marriage regarding her with less than affection, and those by blood referred to as appearing to her ". . . like buzzards flying around waiting for me to die."[13]

[13]Quotations in this paragraph were from the Reverend Fay Lincoln Gemmell, in a converstation of 13 June, 1991, at his home in Keene, NH. He had done chores and errands for Constance while attending theology school in Boston during the years 1944-46.

III

Winning International
Recognition

STONER
UATE IS
HAPLAIN
alumni of
College of
announce-
appointment
army with
of Rev. Fr.
uated from
in the same
Clarence T.
assigned to
San Fran-
Texas, col-
of George
Boone. He
niversity of
r which he
and high
ate work at
izona three
h the medi-
World War.
inishing his
iver he had
diocese and
Daniel J.
e has been
S. veterans
nent histor-
1928. He
I AIMAC

Lowell Observatory Find of New Planet Scientific Achievement of Century

Ranking as of first importance among the scientific achievements of the century, the discovery by the astronomers at Lowell Observatory of a heretofore unknown trans-Neptune celestial body has centered the attention of the world upon this Flagstaff institution.

The first release of news of the discovery was made by the authorities at Harvard college who had been informed through Roger Lowell Putnam who had received the official announcement from Lowell Observatory; only a few hours afterwards the people on Mars Hill and the news representatives in Flagstaff, were objects of bombardments from all quarters by wire and phone mesages, asking details of the discovery and pictures of the newly found celestial body.

News Startles World

This demand for authoritative statements and photographs yet continues, from such papers as have not been served by the news agencies. Besides matter released directly from the Observatory,

WONG
HAD 3
REST

In accor
tom, but a
Flagstaff,
opening of
chen by W
Wong Jun
The occa
true Chine
tom of the
observe im
in an elabo
staff peop
of the hos
and his fa
sions, but r
affair as t
in Flagsta
sons, by
present at
noon, 6 o'c
The rest
modern in
strictly up
been instal
ers sent b
business fir
Guests w
by Mr. an
their son,
Mrs. Fran
(Contin

Momentous news hits the Flagstaff press on March 21, 1930.

7

The Instrumental Man[1]
1894 – 1956

N OT MANY people are privi-
leged to read their own obituary,
but Stanley Sykes read his in a
Wickenburg, Arizona, cafe at his first sit-down meal three days after the
dam at Walnut Grove gave way in the early morning of 22 February,
1890. The flood took an estimated eighty or more lives of those living and
working along the Hassayampa River below the dam site. Stanley was
reported drowned in the disaster.[2]

The idea for a dam near Walnut Grove, just south of the old town of
Wagoner, was conceived by Wells H. Bates, an influential local businessman, who went east in 1885 to enlist capital. Frederick William
Dillingham (1860-1918), a New York cotton merchant, was convinced
and, with others, advanced $350,000 to organize the Walnut Grove Water
Storage Company to provide water for placer mining and irrigation of
flat farmland some twelve miles downstream, below Fool's Gulch where
the valley broadens out. Professor William E. Blake of Yale was in charge
of construction when the project began in August 1886. He built a narrow, mining car railway to haul granite from the company's quarry the
short distance to the construction site and established his own sawmill to

---

[1] This chapter was compiled by Dr. Henry Lee Giclas, a staff member of Lowell Observatory for longer than he cares to admit.
[2] For years, Stanley carried in his wallet the clipping from the *Phoenix Daily Herald* of 25
February, 1890. However, the story was retracted in the *Prescott Courier* under date of 5
March in a column headed: "SAVED - Those who were fortunate enough to escape with
their lives."

The newly completed face of the Walnut Grove dam in 1886

produce lumber and flume-building material. It was reported that these innovations saved over $50,000 in construction costs.

The news reports of the day varied widely in their notes on the dimensions of the dam. It was reported to be from 60 to as much as 135 feet thick at the base; from 200 to 400 feet wide at the top; and 60 to 110 feet high. The closest agreement was that it was 10 to 12 feet thick at the top. Later measurements at the site confirmed a width at the top of 200 feet, and a height of about 84 feet above water level. Since the depth of silt above bedrock at the narrowest point between the two abutments is not known, the total height of the dam must have been somewhat greater. The distance between the abutments is 90 feet, and one can find evidence of a cemented masonry base 28 feet thick on the west side. It is an excellent location and the dam undoubtedly would have been in existence a century later if it had been properly engineered and built.

There were two iron pipes of 22 inches diameter built into the bottom of the dam for controlled release of water. Still in evidence is the 60-inch diameter diversion tunnel blasted through the east abutment. A waste-way 30 feet wide and 8 feet deep was provided on the west side. This, except for debris sloughed in from the uphill side, is still visible. The dam was faced on both sides with hand-laid masonry blocks cemented together. These were less than one foot thick and no one stone weighed more than 150 pounds. The core of the dam was filled with gravel, quarry spalls and sand.

The lake impounded behind the dam in 1887

It was completed in June 1888 and backed up a mile-long lake, along the shores of which many Phoenix residents promptly built summer homes.

After completion of the dam, the construction crew was moved 12 miles downstream to a camp 2 miles below Slim Jim Creek, where they began construction of a smaller diversion dam 57 feet high, 200 feet wide and 60 feet thick at the base. According to the *Prescott Miner*,[3] it was ". . . built of rock, with a good substantial skin of lumber on its upper face, which is caulked perfectly water tight." By November 1889 the first mile of a 6-mile, 30 by 42 inch, wooden flume was completed to carry water to the company's gold-bearing gravel beds below Fool's Gulch. More than one hundred people were living and working at this construction site, and Colonel Alexander Oswald Brodie (1849-1918), later a territorial governor of Arizona, was the engineer in charge.

The winter of 1889-90 was unusually wet. The Bradshaw and the Weaver mountains, parts of which drain into the Hassayampa, were covered with a blanket of heavy, wet snow. The lake was already full when warm rains came, sending additional great quantities of water into the river above. With the water came brush, trees and logs which further complicated matters. Since the spillway gates were made of wood, they had swollen so much that they could not be opened. Thus the water level rose higher than the dam was designed to take.

[3]Under date of Wednesday, 13 November, 1889.

During the afternoon of 21 February, 1890 dam superintendent Thomas Brown became apprehensive and despatched riders downstream to warn people of a possible disaster. The riders never made it! Water started pouring over the top of the dam, washing out the loose fill between the rock veneer of its faces. About 2:00 A.M. on 22 February, the dam gave way with a thunderous roar, sending a wild wall of water, logs and boulders, some as large as a house, down the canyon below.

Sykes, with William Stout and "Long John" Halford, was placer mining for gold, three miles below the lower dam. Their campsite was on the inside curve of a bend in the stream, to which placement he attributed their survival. It was a rainy, moonless night, and they were trying to sleep in a tent a few hundred feet up the slope from the river. They heard the roar of approaching water and instantly realized what had probably happened. They burst out of the tent in their underwear and scrambled up the hillside over rocks, cactus and bushes in their stocking feet. One of them had the presence of mind to drag along the blanket he had been sleeping under. The water touched them before they reached high ground and they escaped being washed away only because the flood-wave hit the outside bank of the curve first, and in the moments it took to rebound to the inside of the curve, they made it to safety.

Stanley told of standing exhausted in the rain above the flood watching water that had a fluorescent glow—a phenomenon also noted by others who survived. Their feet were bruised and stuck with cactus spines. The first order of business was to pick out the spines as best they could in the dark. Then they cleared a short path among the rocks to walk back and forth on, so as to keep warm in the chilly rain. They tore the blanket in strips, wrapping their sore feet and walked to keep from freezing until daylight came.

In the excitement one of Stanley's partners did not realize he could not open one of his hands. When it became light enough to look at it, Stout found he had grabbed a cholla (cactus) ball while reaching for an assist up the canyon slope, and some thorns had gone clear through his hand. Stanley had to make the man lie flat on his back while he knelt on the arm and pried the hand open with sticks so as to extract as much of the cactus as possible.

When daylight came, they beheld a scene of utter destruction. They had lost everything except their underwear and the blanket. The view downstream showed dead horses and human bodies lodged in the trees at the high-water line. The sides of the canyon were scoured clean. The first body encountered in picking their way downstream was that of an

After the washout, February, 1890

Oriental wearing shoes, which were promptly removed and made use of—as were his torn overalls. Further on they found a corpse with better clothing and another quick exchange bettered their condition.

About this time, survivors of the flood from above the upper dam came along. They were on horseback, equipped with tools and trying to identify the dead. They were constructing makeshift coffins from the debris and burying the bodies as they came to them. Soon the meagre supply of suitable material ran out and it was necessary to bury the bodies in the sand and cover the graves with rocks. Two days later, Stanley and his partners reached Wickenburg where, in the Magnolia Brewery and Saloon, Stanley read his own obituary! By this time, however, Stout's hand had swollen as large as a ham. With relief money given them from collections taken in Prescott and Phoenix, they purchased some old nags and rode to Phoenix to see a doctor.

Stanley Sykes was born in South Kensington, London, England, on 18 April, 1865. His father, Godfrey Glenton Sykes, born in 1842 at Malton in Yorkshire, was a well-known professional, designated by command as artist to Her Majesty, Queen Victoria. At the South Kensington Museum there is a Sykes room where his paintings are still on display. Among the memorabilia in the family's possession, in addition to a few of his paintings, is a crayon holder given Stanley's older brother, Godfrey, by his friend, Rudyard Kipling. Stanley attended a private school in South Kensington and received his mechanical training at Finsbury Technical School in London. He then lived in Wolverhampton, near Croydon, until 1884 when he came to the United States with his older brother. Stanley later combined this educational background and inherited artistic ability to become an outstanding instrument maker.

After arriving in New York the brothers spent some months at various odd jobs, saw the recently completed Brooklyn Bridge, and then began a migration to western Kansas. They were inspired from reading about western America in Mayne Reid's *The Headless Horseman*. The story was set on the rolling prairies of the Southwest which became the ultimate goal of the English wanderers.

Later, through a friend for whom Godfrey and Stanley built a house at Garden City, Kansas, they were asked to join in the development of the HCC cattle ranch in the basin of White Woman Creek, some fifty miles to the north. A year later, the same character who sold the land to the owner of the HCC, sold other nearby land to some lady prohibitionists, to the great dismay of the brothers who were operating the ranch for the owner. Then some dirt farmers moved in, the prairies became semi-civilized, and the open range began to disappear. The days of the HCC camp were obviously numbered. Fortunately the person hired by the owner to effect its liquidation was a horse lover as well as a horse trader. For a very nominal sum he allowed the Sykes brothers to purchase their favorite mounts from the remuda. Godfrey described in loving detail the names and temperament of each of the four horses they purchased from the HCC, which they had used during their two years of employment.[4]

After a quick visit back to England, the brothers returned to the West and began a further search for new frontiers. Continuing westward, they left the HCC camp on 14 September, 1886. They took two days to get to Garden City, and two more to reach Ganada, about fourteen miles west of the Kansas-Colorado line. On 21 September they arrived at La Junta, where, ". . . at an indifferent restaurant doing his best at eating a tough steak," they met ". . . a very remarkable old gentleman to whom fighting, in all of its more acute phases, had been more or less a lifetime occupation." It was "Uncle" Dick Wooton, who had built the famous toll road over Raton Pass.

Leaving La Junta they began to see mountains to the southwest, and

[4]See his book, *A Westerly Trend*, published by the Arizona Historical Society, Tucson, in 1944, page 119.
[5]Ibid., page 114.

then traveled thirty-seven miles up the "Picket Wire," their name for the Purgatoire River (El Rio de Las Animas Perdidas en Purgatorio). On 25 September they arrived at Raton, New Mexico, the next day they went on thirty-five miles to Stringer, and the following day another thirty to Wagon Mound. On 29 September, they stopped at Las Vegas Hot Springs for a two-day rest and recuperation. At that time these hot springs were renowned in the Southwest and featured hot mud baths in wooden tubs.[6]

On 2 October they moved on to reach Los Cerrillos and the next day made it to Golden, a little town north of La Madera. Godfrey's comment was: "camped in a rain storm at Golden that was reported to be 25 miles northwest of Albuquerque, but in reality was 100 miles from nowhere— also lost Gotch [Stanley's favorite horse] but found him again." On 4 October they ". . . crossed a big flat at least 1200 miles in width" and arrived in Albuquerque. They "rested" the horses at Trimble's stable and stayed at the Girard House until 9 October. Here they decided to continue on west instead of turning south to southern New Mexico and Arizona.

On 14 October they reached Coolidge, the next day were in Gallup and by the 16th they had passed through Manuelita into the Arizona Territory. Their map showed very little detail of the terrain north and west of the little town of Holbrook on the Atlantic & Pacific Railroad. It showed burning coal beds, waterfalls and volcanic craters somewhat sparsely distributed, and a great blank area labelled "Indian Country" to the north. Intrigued, the brothers spent the next few weeks exploring.

Eventually they drifted towards some high and attractive-looking mountains and finally arrived at the town of Flagstaff. As winter was approaching and they liked the area, they sought a location that would provide feed for their horses. After a bit of inquiry they found a place near the edge of the cedars at about 6000 feet elevation that might be developed into a cow camp. It was close to the blank space on their map and appeared to be unclaimed, suiting their desire for privacy.

Water has always been of great importance in Arizona, and this site contained two natural pools of water in the bottom of a narrow basaltic canyon through which ran a large wash. It had already been named Turkey Tanks. The Sykes brothers built a log cabin with a native stone fireplace near the edge of the canyon and arranged a bucket on a cable so it could be dropped down into the nearest pool and pulled back with a windlass. Stanley said it was the first place they had with running water.

In keeping with the name of the place they designed a turkey track

[6] The Montezuma Hotel at Las Vegas, rebuilt after a fire in 1884, later became a dormitory for the Armand Hammer United World College.

The interior of the Sykes brothers' cabin at Turkey Tanks, 1900

brand (much like the modern highway symbol for a jet airport) and ". . . began to accumulate a breeding herd by picking up odd bunches of cattle as opportunities arose."[7] The brothers did purchase some breeding stock, but in those days, in addition to each cattle ranch having its own brand, many animals were marked with a distinctive ear notch or notches. The practice of ear marking was most useful when, at roundup time, it was much easier to cut out each owner's cattle from their ear marks rather than try to read brands, many of which became obscured by growth of hair. It was considered quite ethical for anyone finding an unbranded calf, to immediately clip his ear mark on it, thus becoming the owner. "Slick ears" became a major ingredient of the turkey track brand.

Their nearest neighbor to the north was Bill Roden who ran a herd near the Grand Falls area. He stopped by Turkey Tanks to pass the time of day and they were sitting out on the porch when a flock of turkeys

[7]Op. cit., page 163.

came by. Stanley reached for his six-gun, aimed, and blew the head off one bird at a considerable distance. It was a one in a thousand lucky shot and Roden was clearly impressed. Stanley, with complete nonchalance, remarked, "I always shoot them in the head; it doesn't tear up so much of the meat." Thereafter, he carefully avoided any further demonstration of his shooting ability in Roden's presence.

Later that year, a bedraggled man carrying a lot of strange gear walked into the Turkey Tanks camp. When he spoke, it was clear he was no ordinary wanderer or hobo; he turned out to be thirty-six-year-old Daniel Trembly MacDougall, then laboratory director of the New York Botanical Garden. His guide had made off with his horses and most of his food several days prior, leaving him stranded on foot. The Sykes brothers took him in and started a lasting friendship which blossomed fifteen years later when MacDougall founded the Desert Botanical Laboratory at Tucson and Godfrey became his associate.

In 1892 the Sykes brothers built a one-room town house at the rear of some lots on Dale and Humphreys streets in Flagstaff. They would hobble their horses in the yard, but almost invariably the critters got away. As it was considered a disgrace to walk to the center of town, they often spent much more time retrieving their horses and saddling them than it would have taken to walk the few blocks into town. This shack later became Stanley's private workshop behind the house he bought from his brother and in which Stanley lived for fifty years. It contained the old foot-pedal lathe, hand-operated drill press, and other mechanical memorabilia of their "Make and Mend" shop.

During the early 1890s the cattle market became so depressed the Sykes brothers decided to try other lines of endeavor. Leaving a trusty cowhand in charge of their camp, they began to utilize their engineering capabilities as operators of a bicycle repair shop in a building on Aspen Avenue, just east of the present-day Monte Vista Hotel. This soon branched out as "Makers and Menders of Anything," an enterprise that became the connection through which they met Percival Lowell when the first 18-inch refracting telescope was installed on Mars Hill in 1894.

The Sykes brothers were intellectuals of the time. They were interested in anything new, and had a way of attracting scientific people by being good listeners. One notable result was that what few well-educated people there were around Flagstaff in those early days tended to meet at the bicycle repair shop to marvel at the mechanical parts the brothers could turn out on their primitive machine tools and forge. This informal group became known as the "Busy Bees" and among their regulars were A. E.

The Grand Falls of the Little Colorado River in spate, 1930

Some of the Busy Bees at Sykes' bicycle shop in 1898. From left are: A.E. Douglass, Balser Hock, Godfrey Sykes, and David Babbit.

Douglass, Dr. MacDougall, postmaster Charlie Stemmer, news store owner Balzer Hock and the Babbitt brothers. They brought with them the growing crowd of itinerant geologists, ethnologists and anthropologists flocking to the scientifically intriguing Arizona Territory.

In 1895 both brothers got married. Godfrey took his bride, an English girl he had known for years, to the trading post he occasionally operated for Tom Keam, in the canyon east of Flagstaff. Stanley married Beatrice Switzer, daughter of a Flagstaff pioneer and settled in at his in-laws lumber mill a few miles west of town.

The next year Percival Lowell engaged the Sykes brothers to build a 40-foot diameter telescope dome to house the 24-inch refracting telescope being completed by Alvan Clark & Sons. They were given no plans for the dome and Lowell was a bit put out when Godfrey asked him what its diameter should be. Nevertheless, it was built on their downtown lot next to the house Godfrey had under construction. After all had been tested on Mars Hill, both telescope and dome were disassembled, loaded on flatcars and shipped off to Mexico.

The Sykes brothers working on the interior of the 24-inch dome

When the Mexican observations were completed, the dome was returned to Flagstaff, where the brothers reerected it at the original location of the 18-inch telescope on Mars Hill. Back in its permanent location the dome was kept round with a steel band that had turnbuckles which could be adjusted to keep everything aligned so it would rotate properly. When it would warp out of round in later years, Stanley always complained that, "those damned peons in Mexico took our dome apart with axes, no wonder it will never stay round." (Since 1966, both dome and telescope have been Registered National Historical Landmarks, though continuing in service. Its turnbuckle unturned since its master's death, with an occasional squeak and grunt, Stanley's dome continues to turn on command, perhaps aided by having its foundations redone in masonry in 1937.)

In 1904 the brothers liquidated the livestock of the turkey-track brand, as Godfrey was advised by the doctor to move his wife to a sea level area due to a heart condition. The older brother then moved to San Diego whence he concentrated on his study of the Colorado River delta. Stanley remained in Flagstaff where, because of his growing reputation as a mill-wright and mechanic, he was increasingly called upon by the Lowell Observatory to install new equipment or repair old.

Percival Lowell always had a great respect for Stanley Sykes; when he came to the shop, he would always knock on the door and wait to be admitted. If machines were running, he stood outside until they stopped before knocking, and then would always remove his hat when entering. They were good personal friends and Lowell often asked Sykes to accompany him on his exploring trips.

The instrument shop at Lowell Observatory during Sykes's tenure consisted of a 13-inch swing radius thread-cutting lathe that had already become eccentric in 1910, a drill press and a shaper. Precision milling machines, flat grinders and other modern machine tools were unheard of at the time. In 1928 Roger Putnam, then the Trustee, donated a small 4½-inch swing radius lathe from the Package Machinery Company. With these crude tools, his mechanical training at Finsbury Tech, and the creative ability of an artist, Stanley produced most of the sophisticated, high-precision, state-of-the-art instruments used by the astronomers at the Observatory in their research over the years. Early in his career he made

Driving clock made by Stanley Sykes in the Lowell shop in 1900 for the 24-inch telescope

the famous fast ratio spectrographic cameras that V. M. Slipher used to determine the high recessional velocities of the spiral nebulae (the first observational evidence of the expanding Universe). These lenses had an F-ratio of 1:1, which meant the cone of light converged to a focus at a 90° angle so the emulsion plate had to be maintained within a few thousandths of an inch. At the other end of the scale, Stanley, from his saw milling experience, could set a huge Corliss steam engine. This versatility and willingness to undertake any mechanical problems made him an indispensable member of the Observatory staff.

The Observatory's screw-cutting lathe was not capable of cutting a period free screw[8] but Stanley produced many of them. He would cut the best screw possible with one cutting, then from an indicator, shift to compensate for the eccentricity of the lathe, and take a final cut. Then he would cast a lead nut on an extra length of the screw, split the nut and use it with abrasive powder, to lap in a perfect screw—a laborious technique learned at Finsbury Tech.

One indispensable accoutrement of the shop was the tobacco dryer. Stanley smoked Mail Pouch tobacco. It was always too wet to burn properly as it came from the package. The dryer consisted of an old oil can from which part of the spout had been cut and a wick inserted. Alcohol was used as a fuel to heat the tobacco emptied into a tin can held on a specially made stand. Stanley always enjoyed opening a new pouch of tobacco. He said it was just like opening a box of Cracker Jack to look for the prize inside. In one package he once found a human thumb, nail intact. Other times he would find buttons, pieces of overalls, and bits of burlap. He always maintained that they never shut down the shredding machine when one of the employees fell into it; there was no telling what might be found in the next package.

Stanley designed and built the mountings for three telescopes at Lowell Observatory; a 15-inch reflector housed in a roll-off roof dome, a 12-inch reflector for the mountain station on Schultz Peak, and the famous 13-inch photographic telescope with which the outermost planet, Pluto, was found in 1930.[9] He built the equipment and participated in two solar eclipse expeditions: one on 8 June, 1918, another on 10 September, 1923, at Ensenada, Baja California. He built the night sky spectrographs that went with the Byrd expedition to Antarctica. In 1936 he made the

[8]A measuring "engine" screw must move the exact amount for each turn of its driving pinion, over its entire length.
[9]Officially known as the "Lawrence Lowell" telescope, after its donor, it is often referred to simply as the "Pluto" telescope.

E.C. Mills and Stanley Sykes in the Mars Hill machine shop in 1927

Wadsworth infrared prism spectrometer for Arthur Adel, with which the latter discovered nitrous oxide and deuterium in the Earth's atmosphere.[10]

A few years later Stanley built a grating spectrometer for Arthur Adel with which the infrared atmospheric transmission and spectrum was mapped. One of the requirements of the instrument was an extremely accurate divided grating table with which to determine the wavelengths of the highly resolved infrared spectra of the atmosphere. There was no such dividing device available at the shop. Sykes solved the problem by nicely machining a hand drive that drove a counter with which the wavelengths were determined. Adel said many years later that a Belgian spectroscopist repeated the measurements of wavelengths with a very sophisticated instrument and could not improve on the measurements made with this one.

Innovation was one of Sykes' great attributes. In the early 1930s Ford was the first automobile to feature an aluminum alloy cylinder head. Since the melting point of this alloy was well below that of iron, Stanley conceived the idea of casting small parts by melting these cylinder heads in a steel plumber's ladle heated in the shop forge. It also required just the

[10]The base for this instrument was made from an old Ford motor flywheel. This history-making spectrometer is now on display among Lowell Observatory's exhibits.

Henry Norris Russell shares a joke with Roger Lowell Putnam at a 1932 astronomical conference on Cape Cod.

right kind of casting sand, with just enough clay to be porous and yet hold its molded shape. To find such material was a great excuse for Sunday expeditions. First along the banks of Oak Creek, then to the Little Colorado River just above Grand Falls where, after a couple of tries, he found just the right combination. A damp handful of this sand squeezed tight would retain its shape. Many lens cells and eyepieces for auxiliary instruments were fabricated from these castings long before aluminum extrusions could be ordered conveniently from a Reynolds Metal catalog.

The spirit of adventure was always with Stanley. In the spring of 1925 he spent a month camping at old familiar places on the Gulf of California. In May of that year he accompanied the senior members of the Observatory staff and the Henry Norris Russell family[11] on a trip to the Indian reservation. In April 1927 he and his wife left Flagstaff in a model T Ford coupe headed for England. They arrived in New York twenty days later, boarded a steamship for Liverpool and visited relatives and his old haunts near London. There were often shorter junkets to interesting places, such as the junction of the Colorado and Little Colorado rivers

[11] A gold medal recipient from the Royal Astronomical Society and others, the 48-year-old Russell was an astrophysicist and research professor as well as director of Princeton's astronomical observatory.

Stanley Sykes, as the author knew him in 1947, repairing an instrument

123 / *The Instrumental Man*

in the Grand Canyon in May 1936. On his seventy-fifth birthday he took the steel-ribbed canvas-covered sailboat he had made to Lake Mead on a fishing trip.

When the administration (now Slipher) building was under construction in 1916, Edward Mills (1866-1958) came to work for Lowell Observatory. He had come west in 1910 and formerly worked for the Fairbanks/Morse Company in St. Johnsbury, Vermont. He was an excellent cabinetmaker and was responsible for most of the fine woodwork around the observatory, including the circular stairways in the rotunda. The Mills family lived only a block away from Stanley, and every day these two men, their lunch boxes in hand, climbed up Mars Hill to work.

It was always a delight to visit the Sykes home. There were many shelves of well-read western history books that Stanley was always generous about sharing with his friends and then discussing. Often, to illustrate a point of conversation, he would break out reciting a Bab Ballad or favorite William S. Gilbert stanza from memory. It was small wonder that the early Flagstaff intellectuals of the "Busy Bee" group enjoyed gathering at the Sykes brothers shop. This carried over to the Observatory, too, where most staff members would bring their technical problems and enjoy Stanley's advice and a bit of his dry, English humor. Once a recently arrived, overbearing, prima donna type (mentioned elsewhere in this narrative by name) was giving Stanley a bad time about some poorly functioning equipment. After a while he replied very simply, "You know, young man, I was at least twice your age before I thought I knew everything."

Lowell Observatory business records show that regular salary checks to Stanley Sykes began in August 1910 and continued for the next forty-six years until his death in April 1956, at the age of ninety-one.

8

And It Just Growed
1894 – 1993

─────────

THE CIVIC leaders of Flagstaff had from the outset promised adequate land suitable for Percival Lowell's observatory. In fulfillment of that commitment, on "this eleventh day of March A.D. 1895 . . ." Nelson G. Layton, Probate Judge of Coconino County and Trustee of the townsite of Flagstaff, ". . . in consideration of the premises aforesaid, and the sum of one dollar . . ." deeded nine lots ". . . being a part of the South half of Section Sixteen, Township Twenty One North, Range Seven East, Gila and Salt River Meridian . . ."(See page 18)

At the bottom of the formal typed deed, Layton noted in longhand:

This deed is issued on the strength of a certain petition signed by Eighty Two citizens of Flagstaff guarantying to said Percival Lowell, a title to said property, in order to that the same may be used as a permanent site for an Observatory, viz: for the aid in scientific observations.

On that same date the document was "Recorded at the request of Prof. A. E. Douglas [sic]" before C. A. Bush, the county recorder, and accompanied in the files by a plat showing the nine parcels in relationship to the rest of the growing town.

As the Observatory haltingly achieved permanent status, its needs for support facilities increased, along with the astronomers' understanding of the suitability of the terrain available on Mars Hill. Responding to the

Flagstaff from Mars Hill, 1912

anticipated need for a larger telescope and more buildings, and for the sum of "Two hundred sixty-two 50/100 ($262.50) dollars, cash in hand" on ". . . this 18th day of June, A.D. 1903," Mayor Thomas E. Pollock signed a quitclaim deed to Percival Lowell for a carefully described further parcel "containing seventy-five (75) acres, more or less." No formal plat was recorded with this transaction, but Percival soon walked the boundary in the company of his recently hired assistant, Carl Lampland, and carefully sketched a map of this newly acquired land.

However, the parcels thus acquired by both gift and purchase from the town were largely along the crest of the hill and included the steep and gullied terrain south and east of the mesa crest. Only part of it was well suited for the best seeing, and Percival felt he needed access to the unoccupied and more level area adjacent to the west where flat terrain induced less air turbulence.[1] He approached friends in Washington and was rewarded on 30 May, 1910, by the passage of H. R. 9304.

An Act Granting certain lands in the Coconino National Forest in Arizona, for observatory purposes.

Be it enacted by the Senate and House of Representatives of the United States of America in Congress assembled, That there be, and hereby is granted to

---

[1]This was part of the area set aside from the public lands as Forest Reserves by President Benjamin Harrison in 1891.

Percival Lowell, his heirs and assigns, section numbered seventeen, in township numbered twenty-one north of range seven east of the Gila and Salt River base meridian, the said tract of land being within the Coconino National Forest in the Territory of Arizona, for observatory purposes in connection with Lowell Observatory: Provided, That in the event of the removal or abandonment of the said observatory or the use of said land by the grantee for other than observatory purposes, the said land shall revert to the United States: Provided further, That the title to the merchantable timber thereon and the right to cut and remove the same in such manner as to preserve the herbage and undergrowth in their natural condition shall remain in the United States.[2]

When a permanent administration building was nearing completion in 1916, negotiations were undertaken with the Forest Service for use of additional land in the southeast corner of Section Sixteen, immediately adjacent to the land already deeded by Act of Congress. This process came to fruition eight months after Percival's death with the signing of a permit by the Supervisor of the Coconino Forest ". . . to use for purposes connected with the Observatory exclusively, 13.8 acres of land . . ." Thus was acquired control, and a measure of the increasingly important protection from the possibility of encroaching lights on land only 93 feet distant from the newly erected Slipher Building and little more than one hundred yards north of Sykes' much-travelled 24-inch telescope dome. A few years later, an additional nine acres was added to round out this parcel.

In order to tidy up land titles and consolidate holdings for ease of administration, in 1972 the Forest Service accepted full title to 60 acres that the Observatory had acquired in Section 30 of Township 22 North and Range 8 East as an exchange for ". . . the 22.5 acres of U. S. National Forest Land on which the Observatory has erected telescopes and a water tank . . ."

This land swap marked the completion of Lowell Observatory's territorial needs on Mars Hill, and along with it momentarily adequate protection from the encroaching lights of the growing city of Flagstaff. But only the Observatory's west flank was really safe from the ever increasing problem of "light pollution" and over the years special projects suggested other temporary observing sites—such as Schultz Peak. However, by 1953

---

[2] After a dispute over his cutting of fenceposts on this property, Lowell sought a clarification of his rights from the Attorney General. James Wilson, Secretary of Agriculture, replied under date of 19 June, 1912: ". . . However, in view of your opinion as to the extent of the rights granted to Prof. Lowell by the Act in question, I have been glad to instruct the Forester that any restrictions upon the use of any of the material on this land by Prof. Lowell which he desires be removed. This action has been taken . . ."

The Clark 24-inch refractor in place in 1902 – where it stayed

the need had become manifest for a permanent, alternative "dark sky" site at a distance from Mars Hill and the ever brightening lights of Flagstaff. Photoelectric study of standard stars needed the greater precision of viewing attainable only with a darker sky. The Observatory was fortunate that there were two other similar searches going on in the area at that time and the various data could be compared. John Hall, later to become Director of Lowell Observatory, and others working for the U. S. Naval Observatory were seeking a location in the Flagstaff area for some major instruments then suffering from the increasingly poor seeing of the Washington (DC) area. The "National Astronomical Observatory" (funded by the National Science Foundation but managed by a consortium of two dozen major universities) was also involved in a locational search which included site evaluation on the high plateau of northern Arizona.[3]

With fairly simple instrumentation, meteorological data was accumulated over a period of months from several potential locations each at some distance from the city of Flagstaff. Then, after careful topographic analysis, a determination was made in 1959 to move the Ohio Wesleyan/Ohio State 69-inch "Perkins"[4] reflector from its almost unusable location at Delaware, Ohio, north of Columbus, to Anderson Mesa,[5] some dozen miles southeast of Mars Hill. This land, under the management of the Forest Service, also became home to Lowell's new 40-inch reflector, the relocated Pluto telescope and the Geological Survey's 31-inch reflector.

Over the years a great variety of structures was erected around the crest of Mars Hill, initially centered about the original location selected by Douglass in the spring of 1894. A few, like domes for two telescopes borrowed for the earlier searches for Planet X, were only temporary. Some structures were dismantled when senility and decrepitude became overriding factors in their continued value, and one burned flat. However, a century after Douglass first set foot on "Site Eleven," Lowell Observatory

[3]The Association of Universities for Research in Astronomy (AURA), an outgrowth of a conference held at Lowell Observatory in late summer of 1953, now manages large facilities at Cerro Tololo in Chile and Sacramento Peak in New Mexico, as well as at its original site on Kitt Peak near Tucson. It has also played a large role in the Space Telescope project.

[4]Hiram Mills Perkins (1822-1924), a long-time faculty member of Ohio Wesleyan University, donated the funding for an observatory and its major component, a large reflecting telescope. See *Popular Astronomy*, Vol. XXXVII, #10, page 553, for a full account of the original installation.

Perkins had done well on the family hog farm as a result of rising meat prices during the Civil War and thereafter was able to enjoy the life he preferred - as a college professor.

[5]Named for James Anderson, of Flagstaff, as is the canyon to the east.

Constructing the oberserver's house in 1894, the beginning of the Baronial Mansion

still made use of twenty-four buildings of various sorts, sizes and shapes, scattered among the ponderosas and malpais. In chronological order of construction these facilities are:

1. The present "Clark Dome" is actually a rebuild on the site of the original Pickering-built dome for the telescopes borrowed to start operations in 1894. Today's structure was set in place in 1896. Its footings were rebuilt in masonry in 1937, which was further rehabilitated in 1990. This nationally recognized historic structure is well mentioned in other parts of this narrative.

2. The "Baronial Mansion" was removed in 1959, after its prior occupants had all died and even the most cursory analysis showed that its rambling, unplanned nature made it unsuitable for modernization. This building was the outgrowth of the original cottage, built late in 1894 to house observers when it became obvious that residency in the hotels downtown was impractical for the largely night-working staff on Mars Hill. Over the years this house became the personal residence of Percival Lowell, with guest quarters for particularly important visitors, his personal secretary, Wrexie Leonard, two household servants, and ultimately for his wife, Constance. The history of this structure is touched upon more fully in other chapters of this narrative.

3. Water, on the dry hilltop, was an immediate necessity. Wells just

The Baronial Mansion from the north in 1907. The area to the immediate right was occupied ten years later by today's Slipher building.

didn't produce in the rocky soils and water had generally been carried to the observers from the more plentiful supplies below. Thus, as soon as arrangements could be made to connect a pipe to the town's water supply system, and install a suitable pump, a stone water tank was built on the highest point of the original Mars Hill property. It was in active use, subject to the vagaries described in Chapter 15, until 1964, when its 12,000 gallon capacity seemed inadequate to assure the growing demand on the hilltop. It was supplemented by a new and larger tank, then taken out of active service.

4. Venus—the cow—needed a home, not only for her apparently annual satellites but for several of her four-footed associates, and so, in 1901, a barn was constructed. Percival took pains to have the lady's quarters placed near an open, grassy area as far south of the water tank as his residence lay to the north. Largely unoccupied and deteriorating in its later years, the barn was taken down in 1943.

5. With the need for housing of additional observers increasing over the years, and Percival's understandable desire for some element of privacy for himself, a second residential building was built in 1902. Largely in the nature of a military post's BOQ (bachelor officers quarters), this building was nearer to Venus's dwelling, but to the east, and became subject to

The V.M. Slipher house in 1925, later modified to the lodge. The gazebo in the left background stood on what was to be the lawn of the house built for William and Ester Baum in 1965.

several additions and alterations over the years. In 1908, at the time of his marriage, much of it was given over to V. M. Slipher and after a minor fire a much larger alteration took place in 1930. When V. M. took up residence downtown, following his retirement in 1954, the building was remodeled once more into "the Lodge," and resumed its original function as quarters for visiting observers and temporary help, until its decrepit and fire-hazardous nature caused it to be withdrawn from all usage in 1990.

6. To indicate his property line, Percival had established a formal gate at the entrance to the Mars Hill property, and under date of 28 August, 1906, he wrote from the Hotel Continental in Paris:

Dear Mr. Slipher:
It seems to me best to have Mr. Sykes build Harry [Hussey, the custodian of the 24" telescope premises, who had recently married] a little lodge,[6] as small as is

[6]Hussey and his brother, Will, both Flagstaff residents, were one-time members of the British merchant marine who, after almost drowning at sea, had decided to settle in as high and dry a location as they could find.

The old gate entry to Lowell Observatory. The gate house built for Harry Hussey, later occupied by E.C. Slipher, is barely visible in the shadows at the left.

consistent with making him perfectly comfortable. It might be placed most advantageously where the entrance gate on the road now is; just above the gate and close to it on the lower side of the road. I would let Mr. Sykes start it at once . . .

In 1919, after Hussey had left Percival's employ, this building became home for E. C. Slipher, and so it stayed for many years. However, Henry Giclas noted that on 14 October, 1936 . . .

The fire that burned this house started from hot ashes Mrs. E.C. Slipher had put out in a box some thirty feet south of the house that caught the dry leaves on fire; and the whole south side of the house was burning before anyone saw it. It was about 11:00 A.M. . . . when the Forest Service lookout on Mt. Elden, Mr. Pratt, called Mrs. E. C. to tell her the house was on fire . . .

The place burned flat and almost nothing was salvaged; but the event got people to thinking about the vulnerability to fire that was endemic to this kind of location. Structures built subsequent to 1936 were all made of more fire resistant materials.

7. When Stanley Sykes took up more or less permanent employment on Mars Hill, as the resident "maker and mender" of everything, his first need was a decent workshop. Percival made provision for this function in a building beside the driveway about equidistant between the 24" dome and Venus' barn, but set off to the west. Added to and altered over the years, the shop was moved elsewhere in 1958, but the building was brought back into service in 1964 as home for the Moon mapping contract with the Areo Chart and Information Center (ACIC) of the U.S. Air Force. After the conclusion of that project in 1970, it came to house the Mars Hill Arts Consortium, an aggregation of visual artists that used the premises for their working studios.

8. In 1908 Percival began planning for a larger telescope, a reflector originally planned to be perhaps as much as seven feet in diameter, which would, for its day, have been a record large mirror. His correspondence on this topic was largely with Carl Lampland, who later came to regard this instrument as almost his own personal toy. However, Dr. Lowell's concerns about location of this new telescope were noted in his letter of 16 March, 1909, to V. M. in which he urged his assistant to take a vacation and then stated:

Thus it seems to me best to sink it in the ground on some hillock to a depth of about 6 feet and then over this to erect a dome about 28 feet in diameter; the excavation need not be to the limit of the whole circumference because the centre of

Constructing the foundation wall for the 40-inch telescope dome in 1909

Assembly of the 40-inch telescope in 1909. Carl Lundin, Jr. is in the white shirt and Eli Giclas, father of Henry, is in the pit.

Carl Lundin, Jr. making adjustments to the pads under the 40-inch mirror, 1910

135 / *And It Just Growed*

The skeleton of the 40-inch dome nears completion in 1909

motion is only about 3 to 4 feet at most from the end of the cell. But there should be a track of concrete laid around it for the dome to run on. When I suggested this idea to Mr. Nolte [Mgr. Alvan Clark & Sons Corp.] he jumped at it as the best solution of the whole matter.

The first thing to do then is for you to choose some spot without cutting down trees where the necessary excavation can be easily made. Why would not the bluff overlooking the mill—where we found the snake's skin—be the place? Or do you think that would be too subject to breeze, jar from the trains, or smoke from the mill plant?

I could leave here any time after April 1st as my lectures[7] end then—so if you will let me know when and how long you wish to be gone I can make my plans accordingly.

I am very anxious that Mr. Lampland should not be requisitioned and even if he were there it would only make my presence the more desirable as someone must keep a tight rein on that spirited noble animal . . .

The exchange of views between Slipher and Lowell continued over the next few weeks while Lowell took counsel with Professor George Willis Ritchey (1864-1945), a distinguished optician and designer of telescopes,

[7]Percival was committed to a series of Lowell Institute lectures at MIT in Boston.

The 40-inch dome shows its strength after heavy snow in 1916

then on the staff of the Carnegie Institute Observatory in Pasadena. In correspondence with both Ritchey and Slipher, Lowell touched on an important aspect of dome construction, noting in summary under date of 20 April, 1906, to Slipher:

> ... He [Ritchey] will I think discover in the course of time that so elaborate and expensive a dome as he proposes is really a spoil-seeing. I said canvas to him—that of course is going a little far. What I mean is more like our own wooden dome. It is most important that we should be able to come to the temperature of the outside air at once since if we do not we simply prolong the agony of doing so and in the second place set up a lot of currents within the dome itself ...

And so it was built in 1909—a dome twenty-eight feet in diameter, well back from the edge of the hill, made of the lightest possible construction consistent with its purpose. And though its most critical contents— the $10,800 mirror—was cracked and disabled in an unfortunate attempt at changing its character from modified Newtonian to Cassegrain[8] focus

---

[8] Everyone knows of Sir Isaac Newton (1642-1727), but Guillaume Cassegrain is more elusive. He was a French physician and designer, largely in the employ of Louis XIV, who submitted his drawings in 1672 for a telescope with the light brought to a focus through a perforation in the primary mirror.

Stanley Sykes helps ready Lowell's Stevens-Duryea for its last trip down Mars Hill in 1938

Revived and beloved by Warwick Eastwood, Lowell's historic vehicle transported Angela Lansbury to the 1992 Rose Bowl game where she tossed the coin to open the game.

in 1959, the dome (somewhat aided by an aluminum skin) was still sturdily shedding the elements more than eighty years later. Later research has established that a turbulence boundary layer seems to exist between twenty and sixty feet above ground, and that the best seeing is obtainable when above that level. In the event, the seeing at this location was seldom very good. However, when Harold Johnson installed exhaust fans in its later years, conditions improved noticeably.

The mirror was ordered by Lowell to be 40 inches in diameter, but Lundin polished the blank of glass out to its full diameter of 42 inches. The cast iron mounting ring, however, covered a portion of the mirror's rim, which was later cut away by Sykes. Thus, this telescope appears in earlier references as a 40-inch diameter and later as a 42-inch.

9. Anticipating the arrival of a specially-made horseless carriage which he had ordered in 1910, Lowell caused a garage to be built that summer, as yet another addition attached to the west end of his Baronial Mansion.

Made by the Stevens-Duryea company of Chicopee Falls, Massachusetts, with an extra-wide rear seat, allegedly to accommodate Mrs. Lowell on her husband's overnight camping trips, upon its arrival Percival's travels around the northern Arizona area took a great surge forward. Sadly, after his death, the big, red, seven-passenger car was mothballed for twenty years and then given away by his widow to some friends in Santa Barbara. After passing through three less appreciative hands, the car was purchased by Warwick Eastwood, of Pasadena, California. He rehabilitated it totally, to its pristine state, entered it in a number of Rose Bowl parades and in 1989 drove it back up Mars Hill to the great joy of the staff members, for most of whom it had been only a pictured memory.[9]

10. Percival Lowell valued all books, and his observatory needed a repository for the accumulation of astronomical journals and associated publications that had been flooding in from all over the world. Thus, in 1910, he built a sturdy separate structure on the forward crest of the hill for housing the Observatory's growing library, almost adjacent to his own dwelling. This was enhanced by one subsequent addition and served well until the burgeoning contents overflowed and were moved to the Slipher Building rotunda in 1959. Even later, and thanks to the generosity of the Transition Fund,[10] the ever-growing contents were transferred to the south wing of the newly built Planetary Research Center.

[9]Henry Giclas had filmed the sad scene when the historic old touring car was driven down Mars Hill for the final time in 1938.
[10]This foundation has aided many valuable causes in the Flagstaff area.

The original library and office under construction in 1910

Percival Lowell contemplates the southerly view from his office

The Slipher building rotunda as the Lowell Observatory library in 1948

11. The principal instigator for Lowell's purchase of the 40-inch telescope was ". . . that spirited noble animal" Carl Otto Lampland. With his marriage to Verna Darby came a need for further housing on Mars Hill, and Lowell supplied it in a house built close below the stone water tank in 1913. After Lampland's death in 1951, this building served as home to other members of the staff, though it no longer housed the astronomer's immense personal library.

12. The centerpiece of today's Lowell Observatory is the Slipher Building, named in 1989 for the brothers whose work, spectrographic, photographic and intellectual became the best known result of Percival Lowell's munificence. Astronomically placed in line with the equinoctial sunrise, but set back so as to preserve one of the native oak trees of which he was so fond,[11] this building, faced with the local volcanic malpais rock, was in his thought process for some years before he finally commissioned his cousin, Guy, to draw up some plans. The structure was initially to be only one-story high, with the prominent, glistening rotunda topping it overall. But, soon after it was completed in 1916, the temporary flat roof leaked so badly that a second story was added with a well pitched roof.

[11] Quercus Gambelii.

This miniature of the Lampland house was tested for the ambience of several locations prior to construction.

Verna Lampland inspects her forthcoming abode in 1913.

The Explorers of Mars Hill / 142

Start of construction for the main administration (Slipher) building in 1915

E.C. Mills (top right) helps pull a precast lintel into place on the rotunda wall.

143 / *And It Just Growed*

The dome of the Slipher building rotunda takes form in 1916.

Laminated sheathing is partially covered by soldered metal, the seams of which leaked minutely for many years after completion.

The one-story Lowell Observatory administration building as completed in 1916

As expanded to a two-story building six years later

145 / *And It Just Growed*

Rededication of the Lowell Observatory administration building in honor of the Slipher brothers in 1989. The second generation included Earl Charles Slipher, Jr., son of E.C.; David Clark Slipher, son of V.M.; William Lowell Putnam, III, current trustee; and Michael Courtney Jenkins Putnam, recently retired trustee.

This building housed all the administrative offices—the director saving for himself the sunrise corner of the ground floor. It soon became fully utilized, the basement being used to hold the Observatory's photographic archives as well as quantities of supplies, darkroom and electronic laboratory. The west half of the second floor was set aside for an apartment, later to be home for many of the staff members during periods of stress or fire elsewhere but primarily devoted to temporary housing for visiting dignitaries.

13. As its physical plant grew, and quarters for resident and visiting astronomers proliferated, so did the need to house vehicles. Thus, in 1919, and north across a wide driveway from Stanley Sykes' machine shop, an eight-stall garage was built.

14. In 1921 a third permanent telescope was installed on Mars Hill—a 21-inch reflector was placed on the axis that had been used for a 16-inch instrument in earlier years. Its building possessed a unique roof structure, designed to roll away totally, thus closely adhering to Percival's earlier expressions of concern regarding massive observatory domes.

Flagstaff from Mars Hill in 1924

Situated only fifty yards west of Lampland's 40-inch dome, the north side of this building has a separate track that holds the roof when the telescope is in use. Seventy years after it was built, both building and contents are still working happily and with continued productivity in Lowell Observatory's ongoing program of solar output studies (see Chapter 11).

15. Built of granite, taken from the same quarry in Quincy, Massachusetts, as the Bunker Hill Monument, the mausoleum holding the earthly remains of Lowell Observatory's founder was completed in 1923. With its own east-facing outlook and with pithy and poignant quotations from his writings, this glass-domed structure reminds every visitor that "... only in places raised above and aloof from men can he profitably persue his search ..."

16. In 1924 a third permanent residence was built near the recently completed administration building. Originally meant to house maintenance and janitorial help, it was remodeled and expanded in 1972 to meet the needs of a newly arrived junior staff member, Robert L. Millis, one day to become director of the Observatory.

17. A one-room, prefabricated house was purchased that same year, initially to provide a lunchroom for Sykes and his occasional assistant. But it languished, since it was more trouble in the winter time to go build

a fire there. In 1947 the little house was enlarged with a bedroom and bath for Charles Kent, the janitor, and later occupied by other support personnel and occasional long-term visitors.

18. On three occasions Lowell Observatory people have made use of locations on the high peaks to the north. In 1908-09 V. M. Slipher sought to substantiate his determination of the presence of water vapor in the Martian atmosphere, made by spectrographic analysis from Mars Hill. In this exercise, he was seeking the same objective as W. W. Campbell in his venture to the summit of Mount Whitney, albeit with perhaps a different motive (See page 163). As he noted in Observatory Bulletin # 84, Slipher found to his dismay that observations

. . . in the dry autumn weather in 1908 from the Observatory and from the Peaks, elevation 12,600 feet, showed that a very much greater height still would have been required to equal good winter conditions at the Observatory, for winter spectrograms at Flagstaff commonly show weaker—often very much weaker— vapor bands than did those made on the mountains.

A few years later, with the completion of a "scenic highway" to an elevation of nearly 11,000 feet, Lowell authorized V. M. to spend $375 for a small structure to hold instruments. However, snowfall made access difficult and atmospheric turbulence on the steep slopes generally tended to

S.L. Boothroyd and V.M. Slipher at Meteor Station on the San Francisco Peaks (altitude 10,500 feet) in 1932

impair the seeing to such an extent that it vitiated whatever advantage the added altitude might have given.[12] Thus, while temporary and special purpose forays were made to Schultz Peak in 1926 and again in 1932 with instruments of various sizes, no sustained effort was ever made at this difficult location.

Of interest, though, in the decision to abandon efforts on the higher location was the development of an aluminizing process for mirrors by Cornell graduate students Henry Crocker Ketcham and Robley Cook Williams[13] that was first applied at Lowell Observatory in 1933. Henry Giclas noted that with the improved characteristics of the telescope, that observations at Flagstaff ". . . could be pushed a few minutes longer with better results. It just wasn't worthwhile going up there, that is why it was abandoned."

19. Among the more famous of Lowell Observatory's buildings is one that was completed in 1929 and is more fully described elsewhere in this volume. The most northerly of the Mars Hill buildings, it is also among the most visited by virtue of being the terminus of the 1990 Pluto Walk, the prime, day-in, day-out, visitor service attraction of Mars Hill. This walk follows the primitive path through the trees that was regularly followed by those engaged in the search for Lowell's "Planet X" in the years after 1929, most notably Clyde Tombaugh. The walk is punctuated by scale images of the Sun and its nine planets, spaced in proper relationship (but on a distance scale reduced to 1/20th of that pertinent to the images). The "Pluto" dome held the "Pluto" telescope from 1929 until its removal to Anderson Mesa in 1970, after which the dome became home to the Ronnie Morgan telescope, acquired as an element of a complex gift and purchase transaction with Ben Owen Morgan of Odessa, Texas.

20. As Lowell Observatory tooled up for its part in the Space Age, the inadequacy of the old machine shop became increasingly apparent. Thus, in 1958 a new, larger and much more fire-resistant building was erected on the high ground near the semi-retired 42-inch dome.

21. In a move to accommodate America's budding space effort, Mars

[12]Giclas recalled one trip ". . . up the Peaks with him . . . when there was lots of snow - every so often we would break through the snow between logs and fall in to our armpits. But V. M. Slipher, 35 years my senior, was always ahead of us 'boys' climbing up the mountain - we puffing and panting and he, disgusted, waiting for us to catch up."

[13]These were both students of Samuel Latimer Boothroyd (1874-1965) who had been a Lowell assistant in 1897-8 and later an assistant on the Alaska Boundary survey. After 1921 Boothroyd was professor of astronomy and geodesy at Cornell University.

The dome for the Lawrence Lowell (Pluto) 13-inch telescope under construction in 1928

The malpais ashlar walls of the dome, nearing completion

Finishing the Pluto dome in 1929

151 / *And It Just Growed*

The Lawrence Lowell (Pluto) telescope in place, 1929

Hill became the focus in 1960 of a Moon mapping effort which involved many of the existing staff and eight people hired especially by the Areo Chart and Information Center for this project. The need for telescope time for this project grew to the point that in 1963 a second but smaller refractor was acquired from Ben Morgan. To provide office space, a modern addition was constructed adjoining the now empty machine shop, once the domain of Stanley Sykes. Known subsequently as the ACIC Building, this later became a computer room, then the repository for much of the subsequent Planetary Patrol photographs of the 1970s and other more or less "dead" storage, before being brought back into use for office space again in 1992. The Planetary Patrol archives were then placed in a special vault in the Planetary Research Center.

22. In 1962, to provide housing for the newly appointed director, John Hall, a new frame residence was constructed near the crest of the "low hill" somewhat east of the Lodge, which had been the residence of the V. M. Slipher family before his retirement. Soon after this was finished, a second (and almost identical) house was built nearby, initially for occupancy by the director of the Planetary Research Center, William Baum.

23. The other 1962 structure, the most westerly of those on Mars Hill,

was originally built to house the Ronnie Morgan 24-inch reflector, used for a Carnegie image tube development. Living quarters and laboratory space were added on its north side to accommodate the people making extended stays on Mars Hill while testing these tubes. The roof over the telescope had the steep-pitched look of an alpine chalet and was hinged at the walls so it could split at the ridge with each side swinging open to totally expose the sky. Seen from outside, the room resembled a large box with the opposing flaps open (in a strong breeze the roof panels did, indeed, flap). Large counterweights protruding from the corners of this roof segment accentuated the alpine appearance of the building which soon acquired the name of "Chalet." The fancy roof was found, however, to induce more turbulence than the more traditional structures elsewhere on Mars Hill. Thus when a spot was found on Anderson Mesa for the Pluto telescope, the Ronnie Morgan reflector was moved into its old dome and the chalet became devoted solely to visitor housing.

24. Early in 1963, the Moon cartographers and commercial artists working for the ACIC needed more observing time than could be provided at the venerable Clark refractor, so Lowell Observatory purchased a 20-inch apochromatic refractor from Ben O. Morgan. A 30-foot diameter dome, 26 feet high with a movable floor was built to house this telescope, near the machine shop at the crest of Mars Hill. In 1984 this long focus refractor was replaced by an 18-inch F1:8 aerial camera of 144-inch focal

William A. Baum at work as director of the Planetary Research Center in 1968

The Planetary Research Center under construction in 1964

The Center was completed early in the summer of 1965.

The Explorers of Mars Hill / 154

Flagstaff from Mars Hill, 1992

length for use as an astrographic camera in deriving accurate positions of stars to be occulted by asteroids.

25. Built largely with the assistance of NASA funds to house the burgeoning staff of Lowell Observatory engaged in preliminary and backup studies for the American Space Program, in 1964 the Planetary Research Center was constructed, with William Baum, who had been an important collaborator in the image tube project, as its first director. This large, modern building added more than 8000 square feet of new office, laboratory and library space, doubling that previously available on Mars Hill. But this same increase also brought on a need to enhance the water storage capability in case of pump failure or fire. Thus, as a companion project, a new, steel tank of 32,000 gallons capacity was situated between the new machine shop and the Pluto Dome. Fire, both as a theoretical possibility and as a reality had been a constant nightmare to Mars Hill personnel and with the new tank, a fire hydrant was placed to allow easy connection in a central Mars Hill location for Flagstaff fire department apparatus.

26. And thus things stood, with a few minor removals and replacements, until plans for the Observatory's centennial observation crystallized around the growing need for a lecture hall and visitors' center that would provide an opportunity to display more effectively the current research activity of Percival Lowell's observatory and thus allow better

exhibition of its historic instrumentation in the Slipher Building rotunda. Adjacent to the record of its past, with the Giclas Lecture Hall,[14] a major part of the Steele[15] Visitors' Center has been designed to show the tools of the astronomer, the current topics of research by the Lowell staff and what the future might hold.

[14]Named in honor of Henry Lee Giclas, whose father helped install Lowell's first 40-inch telescope and whose entire life has been spent in loving and loyal association with Lowell Observatory.
[15]This name is in honor of Arizona's Horace Steele Foundation, the major donor of funding for the Visitors' Center.

9

The Second Trustee
1927 – 1967

XHAUSTED BY his ten-year
struggle with Constance, Guy
Lowell embarked, early in 1927,
for an extended visit to Europe—a visit which was abruptly ended while
at sea near Madeira with his death by stroke on 4 February. He had, however, seen his cousin's estate through the worst and most debilitating
parts of the turmoil caused by the varied interpretations, misinterpreta-
tions, conflicts, court findings and ultimate compromises regarding
Percival's will. The dust had not completely settled, but the framework
for the future course of Lowell Observatory was now firmly in place.
Though Constance and her nephew were to intrude periodically into
observatory affairs for another twenty-seven years, henceforward they
were only a nuisance, no longer a menace, and the residue of the estate
providing income for the continuance of the Observatory, was at last in
solid hands.

In a codicil to his will, dated 19 May, 1920, Guy Lowell named Roger
Lowell Putnam (1893-1972) as his successor, the revelation of which came
as somewhat of a surprise, albeit a welcome one, to the nominee. The second son of Percival's second sister, Elizabeth, was quick, however, to
assume his duties and to enjoy a chance to again exercise his own considerable mathematical talents. He was also immediately in touch with V. M.
Slipher about their mutual expectations, and equally prompt in confirming the status of the distinguished director who had, in fact, been suggested for the position in Percival Lowell's will, despite Lowell's initial statements [see page 45] on the temporary nature of Slipher's employment.

157

There were still a few residual items unsettled from his uncle's estate, requiring conversations with Judge J. E. Jones in Flagstaff, after which the new trustee was finally able to close Percival's office on State Street in Boston, eleven years after its owner had ceased to use it. But there were far more pleasant days ahead for the observatory than in the recent past.

Before he relinquished the responsibilities of the trusteeship to his youngest son, Michael, the second trustee of Lowell Observatory was to guide its affairs for longer than had both Percival and Guy before him, and to see it through a dramatic expansion of its staff and facilities. When Roger Putnam retired from this task, forty years after assuming it, he could count a number of accomplishments:

The fruition of the "Pluto Search" initiated by Percival's calculations and immensely aided by the generosity of Lawrence Lowell;

Final liquidation of the burdensome limitations imposed by the compromise lifetime agreements with Percival's widow;

Acquisition of three new telescopes, and the sharing of one with nine times the light-gathering power of Percival's famous 24-inch Clark refractor;

Establishment of a new "dark sky" viewing site on Anderson Mesa—a dozen miles farther removed from lights of the the growing city of Flagstaff;

Construction of a Planetary Research Center to house the new and growing staff of Lowell Observatory;

Recognition of the work of the Lowell Observatory staff, around the world, as of primary importance in the study of astronomy;

Establishment of an ongoing relationship with the U. S. Weather Bureau, the National Science Foundation, the National Aeronautics and Space Agency, and other organizations, to fund an expanding program of astronomical research.

Roger Putnam's skill and training in mathematics readily enabled him to follow the work of the astronomers, and even to participate to a limited degree. But the trustee's primary duty is to the wise administration of the estate, so as to be able to further the purposes for which the Observatory had been established. Believing in the future, he felt that a wise long-term investment policy is one that positions for future growth regardless of market fluctuations. He also felt that during all the years of indecision and mismanagement of the estate there had been an undue reliance on a "banker's philosophy" in managing the portfolio which, while responsive to immediate needs, is overly attentive to the short range. Therefore, he began to shift the portfolio from an emphasis on bonds and current income to common stocks with long-range growth potential. Within two years, towards the close of 1929, he had achieved what he felt to be a better balance—just as the stock market collapsed.

The stocks he had chosen, however, were prudent, as required by the 1830 ruling of his great-great-grandfather, Massachusetts Supreme Court Justice Samuel Putnam. That "Prudent Man" decision, reached in the case of *Harvard College vs Amory*, saw the enunciation of the legal maxim that ". . . restricts discretion in a client's account to investments only in those securities that a reasonably prudent person seeking reasonable income and preservation of capital might buy for his own investment."[1] Despite the Great Depression and the ensuing deflation of the global economy, the Observatory continued to meet its payroll, and within ten years of this change of investment policy, the estate had improved considerably in net worth.

Lowell Observatory, however, has never been a rich institution. Its various trustees and directors, over its entire existence, have had to pinch pennies at every turn and "panhandle" for improvements in the physical assets necessary to keep up with the evolution of astronomic research. It has not been an easy task, but it has obviously been possible. The endowment, in recent years, has been sufficient to carry only about 30 percent of the annual operating costs—the balance coming from research grants, projects, current donations and generous visitors.

Roger Putnam was a tenth generation Yankee whose ancestry can also be traced back through many more generations in England. In 1640, a year after Percival Lowle had settled in Newbury, John Putnam (1579-1662), from Buckinghamshire, with his wife, Priscilla, three sons, Thomas, Nathaniel and John, and a daughter, Elizabeth, took up land in what was later set off as Danvers, but then known as Salem Village.

A generation later, the Putnam family had one collective moment of debatable glory in 1692 when family members were found on all sides of the celebrated witch accusations and trials that wracked the little settlement.[2] Whatever the root cause of this episode, and many have been advanced, when the dust settled, the progeny of Thomas Putnam's first marriage gradually pulled up their economic stakes and moved away, west, north and south. A few generations later, most of the Putnams

[1] This wording has become the basic investment rule for trustees in cases at law in thirty-eight of the fifty states.
[2] An interesting analysis of this episode can be found in *Salem Possessed: The Social Origins of Witchcraft*, by P. Boyer and S. Nissenbaum; Harvard University Press, 1974.

found in northeastern Massachusetts were descended from Joseph, the only child of the second marriage of Thomas, Jr, to Mary, widow of Nathaniel Veren.

The Roger Lowell Putnam who arrived on Mars Hill on 2 April, 1927, was a sixth generation descendant of Colonel David Putnam (1707-1768), an older brother of General Israel Putnam, one of the great figures of the American Revolution. He was primarily a man of business, president of the Package Machinery Company, a firm his father (William L. Putnam, II) had assembled from various minor components in Springfield, Massachusetts, some twenty-five years earlier. It made machines that wrapped consumer products—candy, cigarettes, chewing gum. When Roger joined it in 1919, as a salesman fresh out of Navy uniform, it had $980,000 in gross receipts. He remained with it, in one capacity or another, for the next fifty-three years and in 1972, the year of his death, gross sales were close to $30,000,000. When he came to Package it had seventy thousand feet of manufacturing space, eight thousand feet of office space and sixty-three employees. The year of his death it employed nine hundred fifty people, and utilized almost six hundred thousand feet of manufacturing and eighty thousand feet of office space.

He was also a man of civic awareness. In 1937 a number of Springfield's civic leaders talked him into seeking the office of mayor—a process that seems not to have been too difficult. He ran on the slogan "A Business Man for a Business Administration," reflecting his concept that local government had very few policy aspects, but a great need for sound and efficient administration. As a further sign of how times have changed since 1937, Roger Putnam's total campaign costs were $2845, and the mayor's annual salary was the magnificent sum of $4000.

However, he never completed his third term as mayor. Early in 1943, with his two older sons already in uniform, Roger Putnam returned to active duty with his commanding officer from World War I. His compilation of gunnery tables for the newly commissioned USS *Mississippi* in 1918 had been so impressive that Commander, now Vice-Admiral Alan Kirk asked him to take on the task of designing special rocket launching vessels for use during the Normandy Invasion.

Thus it was that Roger Putnam begged for tranquility from Keith Kane, the still troublesome lawyer/nephew of his aunt-by-marriage, Constance, at least for the duration of his absence from the country, and headed overseas for England.[3] In 1944, when his services were no longer

[3]He did not get it and the petty harassment continued despite the general wartime willingness to let such matters await the conclusion of global hostilities.

in demand for war, President Roosevelt asked him to set up and manage the Office of Contract Settlement, to arrange a graceful and non-disruptive termination for the many thousands of "until further notice" wartime procurement contracts, the results of which had enabled the "Arsenal of Democracy" to swamp its enemies during World War II. A further stint of federal service came during the Korean War when President Truman tapped him for the distasteful job of Economic Stabilizer.[4]

Early in his tenure on Mars Hill, the second trustee of Lowell Observatory enjoyed a privilege not given to many—indeed only two other identifiable persons in recorded history have had the opportunity to name a celestial "wanderer."[5] It was a high point of his life, the steps leading up to which are told elsewhere in this volume and recounted in scholarly detail by one-time observatory staff member, Bill Hoyt.[6]

After the first announcement of the discovery of Lowell's Planet X had excited scientists and laymen around the world, the challenge remained to settle on a suitable name for this latest celestial discovery. The trustee's mother suggested "Cronus," which met many criteria except that the distrusted Dr. See had used it in a prior prediction of his own. The trustee's widowed aunt, Constance, urged the name "Lowell" in a letter of 16 March, 1930, because ". . . This morning about five I awakened with a strong force taking possession of my attention . . ." Near the conclusion of a subsequent article in *The Scientific Monthly,*[7] Trustee Roger Putnam and Director V. M. Slipher wrote:

. . . It is a coincidence worthy of remark that March 13 is also the anniversary of Sir William Herschel's discovery of the planet Uranus.

[4]See *A Yankee Image*, the Life and Times of Roger Lowell Putnam, by William Lowell Putnam, III; West Kennebunk, Maine, 1991.
[5]The English word *planet* comes from the Greek *planetes*, meaning "wanderer." The ancients, of course, had no knowledge of the countless asteroids that have been identified and named in more recent years.
The planets from Mercury through Saturn were known and named for centuries. Herschel, however, named his discovery "George's star" in hopes of being named Astronomer Royal. While he soon got the cherished job, fellow scientists were unenthusiastic about the name and, with time, the more appropriate Uranus became accepted. In the case of Neptune, Gallic chutzpah attempted to impose the name "Le Verrier"; but, again, world scientific opinion settled on a more traditional name. In selecting Pluto, Roger Putnam sought, with complete success, both to memorialize Percival Lowell and conform with traditional nomenclature.
[6]See W. G. Hoyt: *Planets X and Pluto*; University of Arizona Press, Tucson, 1980.
[7]January, 1932; Vol. XXXIV, pp. 5 - 23; *Searching out Pluto - Lowell's Trans-Neptunian Planet X.*

In passing, we can only remark upon the tremendous, instant, and persistent demand from the public press for information. We did not feel at that time that we had much ready to give out, and were, besides, too much occupied in following the urgent phases of the work to do more than give a few brief comments. The number of telegrams, cablegrams and letters received at the observatory, as well as the large amount of space given the finding of the new planet by the newspapers and magazines, is clear proof of the wide-spread interest in astronomy today . . .

. . . It was also suggested that the planet be named Pluto, and carry the symbol PL. Many people have asked why we chose the name Pluto. We considered carefully the numerous suggestions offered by many serious minded people. Minerva received a plurality, nearly half the letters suggesting this name. We, too, would have liked to use it if it had not been preempted long since by one of the earlier asteroids.

A great many other names were suggested, but the two most popular ones were Cronus and Pluto. Mythologically, there were very good reasons for both. But Pluto is much better known, and his two brothers, Jupiter and Neptune, are already in the heavens. When these three brothers drew lots for the partition of the world, Pluto chose the outer regions. What then could be more appropriate than his now being in these outer regions, and being in the heavens with his two brothers?[8]

The symbols of all the existing planets, except that of Neptune, are so conventionalized and similar that we felt that the proper symbol for Pluto would be the letters "PL" in the form of a monogram which would be easy to write, and to the layman would bring Pluto to mind. The fact that these were Percival Lowell's initials added weight to the thought.

A few astronomers were inclined to question the nature of the object because of the high eccentricity of our first orbit, but this is only temporary. Not long afterwards better orbits were possible because they could be based on much more favorably timed observations and several such orbits were soon forthcoming. These left little doubt as to the nature of the object . . .

It will now be of interest to see how the recent orbits, determined from the more extended path of the planet and quite reliable, check with what Dr. Lowell had predicted. Dr. A. C. D. Crommelin, of England, whose life work has been largely in the orbit field, wrote on May 20:

"I send circular with the orbit that I deduced for the new planet, assuming that it was photographed at Uccle[9] in 1927. It is in good accord with Lowell's forecast:

[8]This passage is pure Putnamese, the fruit of his classical education.
[9]The site of the Royal Belgian Observatory, near Brussels.

"He (Lowell) gave	My (Crommelin) value
Long. of Perih. 205°	216°
Perih. passage 1991	1884.9
Period 282 yrs.	265
Eccentricity .202	.287
Inclination 10°	17°

"It is very difficult to think that all these points are pure flukes!
"It was not till my orbit was in print that I made the comparison with Lowell's prediction, so that there was no cooking in arriving at the agreements."

Not everything came so easily. After the well deserved encomia surrounding the discovery of Pluto, which followed the scientific community's acknowledgment of the outstanding scientific prowess of the elder Slipher, little more was forthcoming from the work conducted on Mars Hill.

Lampland, in particular, was an increasing disappointment. Noble-spirited he might well be, and a diligent scientist, too, but most of his good works—like those Shakespeare attributed to Caesar—were to be interred with his bones. Lampland seemed to have an inordinate fear of publication—lest he be open to criticism for the slightest flaw in his research or reasoning. In this, he was at a far different pole from the man who had hired him. Percival Lowell throve on questioning any complacency of scientific thought and controversy seemed always to surround his pronouncements.[10] His subordinates, more in keeping with their professional training, resisted the limelight of public attention and tried to conform to the peer courtesy process that Lowell's patrician outlook often led him to feel unnecessary.

But the autocrat was gone and his estate, on which the Observatory relied for its entire sustenance, had been severely depleted by the mismanagement during Constance's inter-regnum. The trustee had to pinch

[10]Of his critics, one of the most outspoken was William Wallace Campbell (1862-1938), later to become president of the University of California but then director of its Lick Observatory. In 1909, he even went to the extent of mounting a special expedition to establish a temporary observatory on the summit of Mount Whitney, high above most of the Earth's atmospheric water vapor, just to debunk V. M. Slipher's announcement of having verified (by spectrographic analysis) the presence of water vapor in the Martian atmosphere. Campbell, like most of the professionals, resented the patrician attitude yet simultaneously popular appeal of Lowell's pronouncements (often referred to as "polemics" by his critics) on the possibility of intelligent life having once existed on Mars. A delightful account of this expedition, "To Climb the Highest Mountain," was written by Campbell's more recent successor as director of Lick, Donald Osterbrock, for the *Journal for the History of Astronomy*, XX(2), pp. 77-97.

every penny twice to keep the place going. Salaries were hardly comparable to those offered by the competition, even in those Depression years and there was little cash to spare for bringing in new blood. Those who came to work on Mars Hill were constantly made aware of the need to improvise and make do. But poverty cut two ways.

With whatever spare cash they could find, the Slipher brothers very astutely invested in local real estate, Lampland in books; lesser staff members had part-time diversions or sought other and better paying positions. As time went by, the extraneous business activities of the director and his brother took up more and more of their time—E. C. even became Mayor of Flagstaff. Lampland groused about these matters to the trustee, but still failed to do his share by publishing the scientifically intriguing results of his work with thermocouples in determining the energy output of distant stars.

Knowing that new blood was needed, the trustee was open to innovative, hopefully inexpensive, ways of enhancing the scientific visibility of Lowell Observatory. Here, however, he found himself hampered by the only too human reactions of the senior staff members to any perceived threat to their scientific prestige or positions.

However, in 1933 Arthur Adel, a twenty-five year-old doctoral graduate student at the University of Michigan, after one failed attempt, arranged a meeting with both the trustee and the director in Chicago.[11] Adel had hoped to confer earlier with Slipher at a meeting of the Astronomical Society of the Pacific in Salt Lake City, but the by now distinguished director was in London delivering the annual George Darwin lecture (on his spectrographic studies of the planets) to the Royal Astronomical Society.

After a bit of horse trading Adel landed a job at the munificent annual salary of $1000. His duties would be initially at Ann Arbor under the supervision of chairman of the university's department of physics, Dr. Harrison McAllister Randall.[12] Back in the laboratory, Adel constructed an absorption cell $22\,^1/_2$ meters long, with a light path that doubled back on itself to attain 45 meters. In order to approximate the atmospheres of the giant planets and Venus, methane and carbon dioxide were introduced into the cell at pressures up to forty earthly atmospheres, and

[11] Adel had been pointed at the opportunity by his mentor in infra-red studies, under whom he had just received approval for his PhD thesis, David Mathias Dennison (1900-1976), a brilliant contributor to theoretical physics.
[12] Randall (1870-1953) was a widely respected authority on the behavior of gasses and the infra-red spectra of elements.

ammonia up to its liquifaction point (about seven atmospheres).

Knowing, from the work of V. M. Slipher, what the spectra of these planets indicated by visual means, Adel was able to vary the mixtures in his pressure cell to exactly duplicate the spectrographic results already obtained at Flagstaff. Thus, he was able to determine what had eluded the observers—the precise combination of chemicals causing the absorption bands that appeared in the plates taken of these planets on Mars Hill. This discovery was sobering and doubtless unwelcomed by the delicate sensitivities of the senior scientists.

The first year's work was so successful and scientifically notable that his contract was extended for a second year—and more importantly—his salary increased to $1200. During the second year, Adel's major research was to investigate the solar-telluric spectrum in the far infrared.[13] At that time it was not known at what wavelength atmospheric transmission of extra terrestrial radiation actually ended, nor what transmissions were involved, nor even the accurate existence of atmospheric absorption bands. The study of infrared radiation, which includes the larger part of the totality of radiation from many stellar bodies was then in its infancy. This investigation was of fundamental importance in order to understand the infrared behavior of the Earth's atmosphere, through which all extra-terrestrial objects were seen.

After a year at Johns Hopkins, Adel took up residence at Flagstaff and continued his experiments, subject to a few annoying manifestations of displeasure from his seniors.[14] Late in 1936 space was made in the east basement rooms for a radiation laboratory and a heliostat installed on the roof above. Sykes built a prism spectrometer and Adel arranged for the delivery of a state-of-the-art thermocouple from the University of Michigan and special selenium filters made for him by Professor Pfund at Johns Hopkins.[15]

With this equipment, Adel completed his definitive portrayal of the rock salt prismatic solar spectrum, a major achievement, and went on to

[13]Telluric lines are induced into the spectrum of all light passing through the gasses of the Earth's atmosphere because of selective absorption.

[14]Working with the spectrograph through which V. M. had noted the enormous radial and other velocities of distant galaxies and within a few weeks of his arrival in Flagstaff, Adel found the carbon dioxide bands in the spectrum of Venus that others had noted earlier on Mount Wilson. He was never allowed to use the 24-inch telescope again.

[15]August Herman Pfund (1879-1948) was the recipient of many medals for his work in optics and the infra-red. Lampland, however, pre-empted the advanced thermocouple and left Adel to work with the older "inkwell" models he had made up for planetary radiometry.

measure and record the thermal emissions of the Moon, thus demonstrating that it radiated as a black body. He then wanted to search beyond the 13 micron "window"[16] for additional opportunities through which astronomers could observe the thermal radiation of extra-terrestrial bodies. However, for this purpose he needed a prism made of potassium bromide, for which the observatory had no funds. Fortunately, C. G. Abbott, Secretary of the Smithsonian,[17] understood the importance of this work and, having learned of his protégé's impasse, was able to provide the necessary prism. Once it was installed, Arthur Adel was able to discover the existence of the now famous 20 micron "window" through which astronomers have since been able to determine the heat budgets of the giant planets and probe the nuclei of distant galaxies.

Following exhaustive investigations of the solar-telluric spectrum with prisms and specially made gratings, it became obvious that the program should be directed towards specific planets. This required a smaller version of the spectrometer for attachment to the now 42-inch reflecting telescope—still Lampland's personal domain. Thus, when the apparatus was constructed, the project came under the control of the senior scientist and Adel was effectively denied useful access. Unfortunately, typical of Lampland's possessive secretiveness in his later years, no further results were forthcoming.

Frustrated by the attitude of his seniors, Adel was understandably discouraged—he was onto a great leap forward in astronomy and simply could not get the increasingly immobile Mars Hill functionaries to view him as an institutional opportunity rather than a personal threat. With the outbreak of World War II, E. C. Slipher, who was on the local draft board, made it clear to Adel that if he did not join the war effort, he would end up in the ranks. Adel thereupon left Mars Hill and spent the war years as a civilian scientist working for the U.S. Navy.

When the war was over, Adel met again with the trustee, who was still in Washington winding up the Office of Contract Settlement. A considerable discussion was held about the future of infrared astronomy in general and Lowell Observatory in particular. Adel hoped this would mean an opportunity for him to return and possibly become director. But, economic reality in the form of renewed difficulties with Constance's nephew, combined with the trustee's personal feelings of loyalty to the men his uncle had put in place to make this dream impossible. Adel was

---

[16] A micron is small - one thousandth of a millimeter.
[17] Charles Greeley Abbott (1872-1973) was most noted for his 1953 determination that there was a correlation between solar output variations and terrestrial weather.

again disappointed, and in retrospect, so was Lowell Observatory. In due time he returned to Flagstaff anyhow, but at Northern Arizona University, where he built the world's first infrared telescope and distinguished himself so much that, in 1984, the mathematical sciences building was named in his honor.

On 18 June, 1932, V. M. Slipher had written to his friend and mentor, John Miller, in which letter he discussed the preparation of equipment for use during a forthcoming eclipse and then responded to a previous letter from Miller that had been largely discouraging of the idea

. . . of trying to secure the large Perkins Reflector for a few years intensive observation work here. My information on the situation at Ohio Wesleyan University I got from Dr. [Frank Elmore] Ross of the Yerkes Obs'y who recently visited us. He also told us of the plan whereby the Yerkes Observatory builds and operates for 30 years (I believe it is for 30) the new Observatory of the University of Texas. I had thought it might be possible to move the Perkins reflector out here and use it profitably if the period could be as long as ten years that the Ohio Wesleyan University might be induced to let it come. Ross was of the opinion that the immediate prospects for the big reflector were not good. He mentioned that [Otto] Struve [also of Yerkes] had been granted permission to come to Delaware [Ohio] to do some observing with the Perkins reflector, but I do not believe he will be able to give much time to that if the Yerkes Observatory takes over the task of building and equipping the Texas Observatory. You probably know all about this plan. I did not mention to Ross my dream that we might hope to work out some plan for having the Perkins instrument here for a time. I have not made any progress toward making an approach to the negotiations thinking it well to let the idea simmer a while. Any other suggestions from you will be most welcome.

And simmer it did—for another quarter century—coming to life again in 1957 when Harold Johnson (about whom more in Chapter 11) stimulated the search for a darker sky site than Mars Hill had become with the growth of Flagstaff. With the approval of the trustee, and the cooperation of several other entities also seeking improved locations, a series of sites were explored. Johnson summed up the options in a letter to John Scoville Hall, under date of 13 December, 1957.[18]

[18]Hall was still at the U. S. Naval Observatory, but about to assume the directorship of Lowell Observatory.

The completed dome for the Lowell Observatory's second 42-inch telescope, later named for John Scoville Hall, in 1960

Retiring trustee Roger Lowell Putnam is surrounded by Lowell Observatory associates who attended the 1967 International Astronomical Union meeting in Prague. From the left are: Kent Ford, Otto Franz, Vera Rubin, William Baum, Ruth Hall, John Hall, Roger Putnam, Bernice Giclas, Henry Giclas, Katherine Kron, Peter Boyce, and Gerald Kron.

It is obvious that Padre Butte is the most accessible and easiest to develop of all these sites. However, it has several disadvantages. It is half a mile south of highway 66, about 22 miles east of town. Lights and dust (from the nearby desert) make it a doubtful, although cheap, site. Seeing tests indicate that it is satisfactory from that standpoint. No fire danger.

A-1 Mountain is 5 miles west of town and 2-3 miles north of highway 66. In the future, lights may bother this site a little. The Forest Service says that the fire danger is by far the greatest on A-1 Mountain of all these sites.

After getting these costs, and considering the Forest Service requirements, we here have felt that Woody Mountain is out.[19]

We here have felt that the Anderson Mesa site (which is on the edge of Anderson Mesa overlooking Lake Mary about one mile this way from the upper

[19]Understandably, the Forest Service insisted on having 360° visibility, day and night, for the fire lookout tower on this crest.

1993 aerial view of Anderson Mesa shows the domes of Lowell Observatory present and future instruments. The Near Earth Asteroid dome is at lower left, the Perkins and Hall domes in the left center, and the National Undergraduate Research dome (31") in the right center with the San Francisco Peaks in the distance. The Optical Interferometer buildings in the lower right are still under construction.

dam) is the best bet. It should have the longest probable life since Lake Mary region is Flagstaff's water shed and development is unlikely. It is about 10 miles (airline) SE of Flagstaff and lights should not be significant. The airport beacon is about the same distance as from the Navy site, but is not visible from the site (it may become visible from an elevated platform, I do not know). There should be no more dust than here at Lowell Observatory. The seeing on several nights was satisfactory. There is negligible fire danger.

Both Anderson Mesa and the Perkins telescope came together at the same time, though not without some intense haggling that even had the trustee writing a three-page letter of complaint to Arthur Fleming, then Secretary of Health, Education and Welfare. Finally, in 1959 an agreement was reached whereby Ohio Wesleyan and Ohio State universities (by now joint custodians of the Perkins premises) bought the idea that V. M.

had been angling for more than a quarter century earlier. With generous aid from the National Science Foundation, the Perkins telescope, at the time fifth largest in the United States, was moved to the site that Lowell Observatory had now developed on Anderson Mesa. Lowell agreed to house the instrument in a suitable dome and thus received half the time available on it; Ohio State sent a series of observers and finally a staff member to live in Flagstaff full-time to service the instrument and make use of its half of the available time.[20] Six years later, additional help from the National Science Foundation enabled the joint proprietors to replace the ancient 69-inch mirror with a larger 72-inch, thus increasing its light-gathering power by 9 percent.

Because of the proprieties of these things (and the delicacy of scientific egos) the telescope is now referred to as: "The Perkins Telescope of Ohio Wesleyan and Ohio State Universities, at the Lowell Observatory." The pang of sending their prized instrument so far from home was eased for the Ohioans by the generosity of Michael Schottland of Martinsville, Virginia, who gave Ohio Wesleyan a 32-inch reflector and several smaller telescopes so that the massive fifty-five foot dome that had been home for the Perkins telescope since 1931 would not look quite so vacant.

[20] It was estimated that the telescope would be four times more useful in Arizona than it had been in Ohio; thus the Ohio astronomers, while appearing to lose half its time, were in fact gaining double.

10

The Search For Planet "X"[1]
1902 – 1930

NOT THAT he had a limited purview relative to astronomy, for Percival Lowell, personally or through his various assistants, investigated a great variety of astronomical questions, but he is primarily remembered for two scientific theses - the "sometime habitability" of Mars, and the existence of a ninth planet in the solar system. Reports and "discoveries" relative to Mars were controversial among the astronomical fraternity, even before Lowell took up the mantle of leadership on "canali" from Schiaparelli. He was undeterred, however, by the shrillness of his opponents on this matter and even less fazed by the vigor of their rebuttals. He almost seemed to savor their antagonism, and it was vigorous.

Lowell's main claim to remembrance by posterity stems from his insistence that there had to be yet another planet out there—somewhere, in the dim and frigid outer reaches of the solar system. In this contention he was among a slightly larger company than when dealing with Mars. The great English astronomer, Sir William Herschel (1738-1822), had picked out Uranus in 1781 by doing a systematic survey of the heavens, thus extending human awareness of the solar system some 30 percent farther from the Sun than Saturn, the outermost previously known planet. In 1841 John Couch Adams (1819-1892) noticed irregularities in the orbit of Uranus that were not explainable by the other gravitational attractions

[1]This chapter was compiled largely by Dr. Gibson Reaves, Professor of Astronomy at the University of Southern California and long-time associate of Lowell Observatory.

then known to be in orbit around the Sun. He calculated what the source of that other force ought to be and submitted his results to Britain's astronomer royal.[2] That worthy, however, considered a search for Adams' planet not a proper task for his attention until after the French astronomer Urbain Jean Joseph Le Verrier (1811-1877) had come to an almost identical conclusion in 1846 and asked the famous German astronomical observer Johann Gottfried Galle (1812-1910), of the Berlin Observatory, to see if it was there.

Galle, assisted by Heinrich d'Arrest, found the new planet after only an hour of searching. It was almost precisely where Le Verrier had told him to look. Neptune was less than one degree away from the mathematically predicted spot! Yet, as time went on, theorists asked the question: "Does the existence of Neptune, in fact, account for all of the discrepancies in Uranus' motion?"

One of the more prominent among these theorists was the younger of the Pickering brothers of Harvard University, the man who had been so helpful in advising Percival Lowell on the placement of his observatory. From 1908, for the next generation, even after the definitive sighting of Pluto, W. H. Pickering intermittently published his latest predictions of positions of hypothetical planets. His bases were graphical analyses of tiny discrepancies in the motions of Uranus, Neptune, and even Jupiter and Saturn.

Lowell, in contradistinction, worked from a much more rigorous theoretical base—more akin to that used by Adams and Le Verrier—and sought to explain what he thought were discrepancies in the motion of Uranus that were not accounted for by Neptune's presence. Lowell felt that the additional gravitational attraction of Neptune alone did not explain the perturbations in the orbit of Uranus, from which perturbations Neptune's very existence had been plotted.

Whenever the clouds of controversy surrounding his Martian opinions eased sufficiently for him to poke his head into the clear, Percival Lowell considered this problem. Finally, in 1905, he set out to get some answers. This was to develop into a massive exercise in longhand and logarithmic computation—much more detailed than that surrounding the prediction of Neptune, for the orbital discrepancies on which it was all based were more than an order of magnitude smaller.[3]

After consultation with the director of the U.S. Naval Observatory,

[2]Sir George Biddell Airy was Astronomer Royal during the years 1835-81.
[3]Massive electronic computers, such as are in common use today, were two generations in the future. Lowell's "computers" used log tables and reams of paper.

Lowell engaged the services of a 1903 MIT honors graduate, Elizabeth Langdon Williams, as his chief "computer" and over the next ten years hired several others to help her.[4] Their work was done largely in Lowell's Boston office, though some aspects of it occurred in Flagstaff, particularly in the later stages of the calculations in 1914.

Though he had bruited his belief in a "Trans-Neptunian planet" in lectures as early as 1902, it was not until he had the first of the computers working on the theoretical aspects of the project, that he ordered his observatory staff to actually look for something. The initial search was conducted with the 24-inch Clark refractor in a narrow band close to the plane of the solar equator. All the then known planets circled in an orbital plane only 3° wide which closely matched that of the Sun's own rotational plane. The observers on Mars Hill reported no success, for though they used Lowell Observatory's finest telescope, its photographic plates covered an area of the sky only 50 arc minutes in diameter and could pick up objects down to only magnitude 13.5 during a one-hour exposure.

The instrument was obviously poorly suited to the necessarily massive photographic survey and comparison, so arrangements were made to borrow a 5-inch lens from Brashear. Throughout much of 1907 this tele-

[4]These included Earl A. Edwards, Thomas B. Gill, Johnson O'Conner and Herbert H. Tucker.

Elizabeth Langdon Williams, Lowell's principal "computer," as Mrs. G.H. Hamilton in 1923

scope was used by J. C. Duncan, E. C. Slipher and K. P. Williams,[5] but these results were equally disappointing.

Further refinements in the calculations were undertaken with the engagement of William Thomas Carrigan[6] from the U.S. Naval Observatory, to once again systematically reduce the residuals in the longitudes of Uranus and Neptune. But his work, though painstaking and extending over many months, was still inconclusive. The arrival in 1911 of a Zeiss "blink" comparator should have aided the search immensely.[7] Unfortunately, by this time, Lowell's attention was again on Mars and the expensive machine languished.

In 1912 the mathematical portion of the search was resumed under the guidance of Miss Williams and her assistants—a two-year program which Lowell now dovetailed into a second observational effort. Much of this second effort was done with a 9-inch Brashear refractor borrowed from Swarthmore College's Sproul Observatory. Again, however, the work was concentrated along the plane of the solar ecliptic, and again nothing was found that resembled the objective of the search. The effort was not barren, however, for on the thousand plates exposed in this search, more than five hundred "new" asteroids were discovered.[8]

With the death of Percival Lowell, and the lengthy controversy over the direction of the Observatory that ensued, all serious efforts to further its founder's second major interest were shelved. But, the challenge was not forgotten.

Henry Giclas took up the story in his 1980 article:[9]

With Lowell's death in 1916 and the uncertainties created by the litigation of his will which followed for several years, very little progress was made on a search. That it was still very much on the minds of the two remaining senior staff members, V. M. Slipher and C. O. Lampland, is seen from the fact that Lampland at the end of World War I inquired, through Elis Strömgren in

---

[5]These were the three Lawrence Fellows of Lowell Observatory, all Indiana boys. By coincidence, certainly not design, this observatory has been blessed by its fixation on employment of Midwestern farm boys.

[6]Carrigan, forty-two years of age, was already well-known for his studies of perturbations in the orbit of Mars.

[7]A comparator allows two plates of the same portion of the sky to be compared through one eyepiece. If anything has moved between the time of one exposure and that of the second, it will appear to blink as first one plate is shown and then the other.

[8]Lowell Observatory has continued its interest in these minor planets with the work being carried on more recently by Dr. Edward Bowell and dovetailing with that of Dr. Eugene and Mrs. Caroline Shoemaker. This team of observers has found and plotted more asteroids than any group in history - by far.

[9]*Icarus*, Vol. 44; pages 7-11.

Copenhagen, if Max Wolf of Heidelberg, Germany, would lease one of the 16-inch Bruce photographic objectives to the Observatory. Also about this time, Guy Lowell, the first appointed Trustee of the Observatory after Lowell's death, inquired of the Reverend Joel Metcalf[10] of Taunton, Massachusetts, considered one of the most skilled optical craftsmen in America, if he would assist the Observatory in securing a suitable photographic objective. Nothing came of these inquiries at the time; but after the death of the Reverend Metcalf in 1925, Guy Lowell, with his own personal funds, purchased from the Metcalf estate three pieces of unfinished glass components for a photographic triplet 13 inches in diameter. Before anything could be done with them, however, Guy Lowell died unexpectedly of a stroke in 1927.

Roger Lowell Putnam, a nephew of Percival Lowell, then became Trustee and finally that year succeeded in ending the more than 10 years of legal litigation which had hamstrung and demoralized the operation of the Observatory since Percival Lowell's death. As his second undertaking, Roger Putnam assisted Slipher and Lampland in converting the Metcalf glass disks into a suitable search telescope. Funds from the Lowell Endowment were very limited, so Roger Putnam appealed to his uncle, A. Lawrence Lowell, the president of Harvard University and Percival's brother, to give the funds necessary to finish a lens and build a telescope mounting and dome. Ten thousand dollars was the amount of the gift from A. Lawrence Lowell. After considerable discussion and negotiation, it was decided to award the contract for making the lens to . . . the experienced optician[s] with Alvan Clark & Sons, the makers of the 24 and 40-inch telescopes, for the sum of $4000.[11] Mr. Putnam then asked Dr. Slipher if he could build the mounting and dome for the balance of the money donated, to which his answer was "yes."

In fact, Slipher anticipated the commitment from Lawrence Lowell and by November of 1927 had already set Stanley Sykes to designing a mounting for the 13-inch lens that Guy Lowell had acquired from Metcalf. On 2 April, 1928, the essential worm wheel drive was ordered from Brown and Sharpe; a week later the site for the dome had been selected. By 20 November, the polar axis for the new telescope was in place, with Stanley's son, Guy, helping his father to complete the task. On

---

[10] Joel Hastings Metcalf (1866-1925) was ordained in 1890 and served at several Unitarian parishes in New England. The recipient of many awards, he was more than an amateur astronomer, discovering 41 asteroids and 6 comets and serving as chairman of the visiting committee to Harvard Observatory.

[11] Alvan Graham Clark had died in 1897, and his successor, Carl Axel Robert Lundin, who had personally installed Lowell's 40-inch reflector in 1909, had died in 1915. But the firm was still highly regarded and under the management of C. A. R. Lundin, Jr.

11 February, 1929, the lens itself was installed and final adjustments soon followed. The people at Clark had indicated that the lens could photograph as much as a 20° field, but not with perfectly crisp images near the edges, thus the final choice was to use a 14 by 17 inch plate which gave a field of 12° by 15° in the sky and a loss of about one magnitude from the center of each plate to the edge.[12]

At the end of 1928 the stage was almost set, with the right props coming into place, and plans could be made to begin a third search for the elusive "Planet X." The new telescope was en route to its dome, its essential drive machinery in hand, and V. M. Slipher now set out to find an appropriate assistant to undertake the tiresome drudgery of systematically photographing the sky—all that could be seen of it from Flagstaff—and then duplicating each plate a second and third time, at intervals of several days. This time lapse was necessary in order to allow the much greater apparent motion of a relatively nearby planet to change its position against the background of the vastly more distant and therefore "fixed" stars.

On 15 January, 1929, Carl Lampland noted that "the young man from Kansas" had arrived. Over the next few months, punctuated by a lengthy

[12]The successful photographic search for Planet X was done down to a magnitude of 16.5 for most of its coverage area.

Clyde Tombaugh at the door of the Pluto dome with a special plate holder used in the search in 1930

return to the family farm in Kansas, twenty-three-year-old Clyde William Tombaugh settled in as lowest man on the Mars Hill totem pole, where he was to stay for the next fifteen years and achieve a notable place in astronomic history.

However, V. M.'s correspondence makes it clear that he did not want someone with high academic credentials for this task, since it was bound to be a boring and tedious one. The earlier searches had been more like rifle shots, aimed at hitting the target precisely where the theoretical calculations had said it should be. This survey, though starting in the most opportune areas, was to be a long project and completely systematic, covering the entire orbital possibilities regardless of where the planet was supposed to be at any precise time. In any case, Slipher picked the right man for this kind of job and carefully delineated his duties. Tombaugh stayed on with the task of inventorying the sky long after Pluto had been found, and in so doing produced a resource of enormous future value both to Lowell Observatory and the nation.[13]

A few test plates having been exposed in the late fall of 1929, the search got under way in earnest early in 1930 and Clyde began his routine nighttime trudges, in the crisp cold of early dawn carrying his precious glass plates from the telescope site down to the darkroom in the Administration (Slipher) Building.[14]

On 1 February, 1930, Lampland noted in his diary that Tombaugh was "... observing and doing janitor work." But, by the 18th of that month, Tombaugh was "... examining X plates." Then things started to move faster. Two nights later, Lampland himself was at the 24-inch refractor, using the more powerful telescope to get a closer look and verify Tombaugh's initial "blink" sighting. On the night of the 25th, he looked at it with the even stronger 40-inch reflector. It seemed that Planet "X" had at last been found. After the decades of hope and work, though at last armed with the proper equipment, success seemed to have come entirely too easily.

A flurry of further checks ensued with other observatories and on 13 March, the seventy-fifth anniversary of the birth of Percival Lowell, Lampland delivered the Lowell Memorial prize at Northern Arizona State Teachers' College and told the assembled student body that on two plates, taken on January 23rd and 29th, the long-sought planet had most assuredly been found and a preliminary orbit determined. There was lit-

[13]Tombaugh, largely self-taught, in time became a respected professional astronomer and teacher.
[14]Fifty years later, this pathway became formalized as Lowell Observatory's notable "Pluto Walk," a scale model of the Solar System, ending at the dome where its outermost member was sighted.

Tombaugh at the Zeiss blink comparator, 1938

tle doubt that Percival Lowell had finally been justified, even though his planet was much fainter than he had predicted.

Slipher's caution, somewhat reminiscent of the uncertainties vented by Douglass a generation earlier, now came to the fore. Lowell's frequently premature announcements having often led to more scientific heat than light, Slipher wanted to be absolutely sure of his facts before allowing any statement that might lead to renewed criticism of that nature. Ironically, Slipher was soon under fire for having taken this extra time to verify his data.

Following the public announcement, and despite skepticism from a few fellow astronomers, almost everyone went back to their photographic archives and it soon became clear that Pluto had been observed elsewhere and in years past. Three years before Tombaugh's photographs, the royal Belgian observatory at Uccle had taken its picture, but failed to make a comparison with which to "blink" it. Even more poignant was the fact that the ninth planet had also been photographed at Lowell Observatory on 19 March and again on 7 April, 1915, while its founder was preparing for the publication of his 106-page memoir on *A Trans-Neptunian Planet*!

Henry Giclas removing the dust cap from the Pluto telescope in 1947

While the furor of Lampland's announcement roiled in the astronomical atmosphere, the heavenly inventory-taking was continued, as Giclas
wrote:

*Suffice it to say that after the excitement of the discovery of Pluto on plate
Nos. 165 and 171 . . . subsided, it was decided to go on with the survey so that it
would be possible to say that there were no additional distant planets brighter
than the search limit of roughly 16.5 magnitude. An additional 1440 plates were
made after the discovery plate for the extended search program. Clyde Tombaugh
continued the blink examination of these plates until about eight-tenths of the
entire sky paralleling both sides of the ecliptic were searched.[15]*

Even then the Lawrence Lowell telescope's days were not over. Giclas,
himself had used it, starting in 1936, to make a series of short exposure
position measurements of comets and longer exposures to provide both
head and tail information on the brighter ones.

In 1955 a proposal was made to the National Science Foundation for
support of a proper motion program using the original "Pluto Search"
plates and duplicating them a quarter century later. The Pluto search had
covered vast heavenly territory and represented a lot of work as well as
being part of an irreplaceable archive. But, to determine the motions that
had occurred in the interim, meant duplicating those plates and then
blinking the new ones with the old.

It became a Giclas project, and was a while getting under way but, in
1957

*. . . a plan was inaugurated to utilize these [Pluto Search] plates as the first
epoch for a proper-motion survey . . . The entire northern hemisphere and one-
quarter of the southern hemisphere were blink-examined independently by two
different observers for stars with motions ≥0.2"/year. The interval between the
two plates blinked at the inception of the program was 28 years, the last few
plates over 47 years. Over 800 pages of proper-motion data have been published
and distributed, mostly in the* Lowell Observatory Bulletin *(Vols. IV—VIII).*

*Special supplementary equipment for processing these data was developed.
The first piece of equipment acquired at the onset was a precision blink built by C.
Ridell of Chicago. At the beginning, proper motion and direction were read
directly from a calibrated screen, and the coordinates were determined with a
theodolite used as an analog device. Later the whole operation was automated so*

[15]Giclas; op. cit. The ongoing "Pluto Search" resulted in the cataloging of 1800 galaxies
and 700 asteroids until its termination in 1945.

that the raw measures made by each observer were reduced, compared for consistency and then combined and printed out in catalog form complete with coordinates in the sky . . .[16]

Yet, even as this manuscript was in preparation, Pluto remains the only planet in our system whose surface features have not been delineated by human spacecraft. So the saga of the discovery of Pluto is not yet over, though, by use of modern techniques, staff members at Lowell Observatory have already begun to map its surface from afar.

For Percival Lowell, the trans-Neptunian planet had been his then unobserved ninth planet, called "X" for the algebraic unknown. Today, though the odds seem even lower, the search continues for an unknown tenth planet—again, and perhaps even more appropriately—"Planet X."

The history of the search for Planet "X" and the subsequent discovery of Pluto at Lowell Observatory in 1930 have been misunderstood by several commentators who are not familiar with the story behind this discovery. Percival Lowell and his successors at Lowell Observatory were by no means the first or the only astronomers searching for a trans-Neptunian planet,[17] yet only the search at Lowell Observatory was successful. Here follow some of the criticisms of the Lowell effort made by various authors, together with a commentary:

1) *Percival Lowell was merely a dilettante, a popularizer of astronomy. He would have been incapable of carrying out the difficult mathematical analysis that led to Pluto's discovery.*

Percival Lowell is more accurately characterized as a Renaissance man than as a dilettante. He was indeed a popularizer of astronomy especially the idea of Mars as the abode of life. Yet he was also an outstandingly competent mathematician, having graduated from Harvard (1876) with honors in that subject. Benjamin Osgood Pierce (1839-1914), Hollis Professor of Mathematics and Natural Philosophy at Harvard, considered him one of the finest mathematicians he had ever known, and expressed his hope that one day Lowell might succeed him in that professorship.[18]

Lowell's most famous work in applied mathematics is his 1915 *Memoir on a Trans-Neptunian Planet*. In this work, from a rigorous analysis of the

[16]Giclas; op. cit.
[17]In 1877-78, David Todd (1855-1939), then of the U. S. Naval Observatory in Washington, undertook the first such endeavor.
[18]See A. Lawrence Lowell, *Biography of Percival Lowell*, New York, 1935, page 6.

still outstanding residuals in the motion in celestial longitude of Uranus—outstanding after allowance for the perturbations of Neptune—and under the assumption that these residuals were caused by still another undiscovered planet, Lowell determined the orbit of such a trans-Neptunian planet. As is well known, Pluto was found as a direct result of the calculations given in Lowell's memoir.

A detailed critique of Lowell's analysis of these perturbations of Uranus was the subject of the 1941 doctoral thesis of Vladimir Kourganoff[19] at the University of Paris. His conclusion was that Lowell's analysis was rigorous and properly carried out.[20]

Even so, it is true that many popularizers of astronomy and the history of astronomy know so little astronomy that they lead the uncritical reader astray. There is reason to be wary.

2) *The Trans-Neptunian planet Percival Lowell predicted could not have caused the perturbations of Uranus on which he based his calculations.*

In fact, Kourganoff showed that such a planet would indeed have caused such perturbations. Such a planet would have had a mass of about 6 or 7 times that of the Earth, and an apparent magnitude of 12 to 13. But Pluto's apparent magnitude was only about 15, leading to speculation from the beginning that it might be smaller and thus less massive than predicted. The mass of Pluto was firmly established in 1978, only after the discovery of its moon, Charon, by 40-year-old James Walter Christy of the U.S. Naval Observatory. The mass of Pluto is far, far too small for it to have been responsible for the perturbations of Uranus that Lowell analyzed! Suspecting this, mathematician E. W. Brown,[21] otherwise quite critical of Lowell's analysis, wrote in 1930: ". . . in so far as it has stimulated a search for an outer planet which has proved successful, one cannot regret its completion and publication."

Lowell's assumption that the outstanding residuals in the motion of Uranus were caused by his trans-Neptunian Planet was incorrect. But, given that assumption, his analysis of those residuals was correct, rigorous, and led directly to the discovery of Pluto.

A question still unanswered today is what caused those residuals? Was it simply observational error? Was it Percival Lowell's method of

---

[19]Kourganoff, 29 years of age at the time, later became professor of astronomy at the University of Paris. His doctoral examiners were Professors Jean Chazy and Ernest Esclangon.
[20]See G. Reaves in *Publications of the Astronomical Society of the Pacific*, Vol. 63 (1951), page 49, for a summary of Kourganoff's research and for detailed references.
[21]Ernest William Brown (1866-1938) is noted for his theory on the motion of the Moon. Quotation is from *Proceedings of the National Academy of Sciences*, Vol. 16 (1930), page 364.

smoothing those observations? Or was it an error in the calculations of general perturbations[22] in the theory of Uranus?[23]

3) *Because Lowell's two predictions were about 180° apart, they were so inaccurate as to be useless.*

When Planet X, the perturbing planet, is closest to Uranus, the perturbed planet, then because the Earth is relatively so close to the Sun, we on Earth will see Planet X and Uranus close together in the sky. At that time, the attraction by Planet X will displace Uranus towards it, and also, by a lesser amount—because it is farther away and more massive—displace the Sun. The result is that the distance from the Sun to Uranus is slightly increased.

When Planet X is farthest from Uranus, thus as seen from the Earth 180° from Uranus, Planet X will displace the Sun towards it, and also, by a lesser amount as it is now much farther away—displace Uranus. So the result in this case also is that the distance from the Sun to Uranus is slightly increased. Thus, whether Planet X is nearest Uranus or farthest from Uranus, its perturbation of the distance from the Sun to Uranus will be almost the same.

Turning this reasoning around, one can begin to appreciate why, given the perturbations of Uranus, one will find two possible positions for Planet X that are about 180° apart. It is quite similar to the mathematical exercise of finding the direction to the Moon from measurements of the twice daily terrestrial tides.

So Lowell's analysis should be expected to yield a double solution, and it did. Lowell found that one solution was very slightly better than the other, so he favored it. And that was where Pluto was discovered. At the very least, it was, as the Smithsonian's Brian Marsden put it in 1941, "an incredible coincidence."[24]

4) *It was pure luck that Lowell's interpretation of the residuals in the motion of Uranus led to an accurate prediction of Pluto's position; Percival Lowell deserved no credit for the discovery of Pluto.*

Two fundamental points contradict this claim. Percival Lowell's prediction—whatever its basis — was, in fact, quite accurate. Percival Lowell

---

[22]In Percival Lowell's time, the motions of planets were represented by formulae consisting of a series of trigonomic functions of the time, a technique usually referred to as general perturbations. Planetary motions are now calculated using step-by-step numerical integration, technically referred to as special perturbations.

[23]The "theory" of any star or planet is the path it would take in its motion through the Universe, taking into account all the known forces acting upon it.

[24]Quoted from Marsden's biography of Percival Lowell in the *Dictionary of Scientific Biography*, Vol. 8; C. C. Gillespie, editor; 1973, page 521.

had unshakable confidence in the correctness of his prediction and inspired others with this confidence together with the will to look for his planet. This *legacy of conviction* was so strong that it carried over from the time of his death in 1916 to the discovery Pluto in 1930.

Clearly, it was not enough merely to predict the position of the trans-Neptunian body. Were it not for Lowell's persisting influence, Pluto would not have been discovered at Lowell Observatory, if at all. Thus Percival Lowell deserves twofold recognition: credit for the prediction of Pluto, and credit for the discovery of Pluto.

5) *Who gets credit for what is not important; what is important is that Pluto was discovered.*

Credit, recognition for achievement, is the foundation for reputation. Any person's good reputation will help that individual gain the support and collaboration necessary for continuation of his or her work. Furthermore, it is often difficult for any scientist to assess the quality and significance of his own work. Thus the researcher is both informed and encouraged by the credit given him by his peers.

Lastly, while some people find their rewards from the work itself, for others the reward is the recognition, the credit; the work is merely a means to that (egocentric) end.

6) *W. H. Pickering's predictions were ill-founded, inaccurate and useless.*

Where Lowell attributed the existence of the "canals" on Mars to Martians, W. H. Pickering[25] attributed changes in the reflectivity of the floor of the lunar crater, Eratosthenes, to swarming hordes of ants.[26] (For the swarming rate of the ants, Pickering cited a paper by Harlow Shapley.)[27] And where Percival Lowell predicted only one planet, by 1930 Pickering had published predictions of the orbital elements for six trans-Neptunian planets. One of these predicted objects, his Planet O of 1919, turned out to be very slightly closer to Pluto's position than Lowell's prediction of Planet X. Further, where Pickering's predicted magnitude for Planet O was 15 and Lowell's prediction for Planet X was 12 to 13, the true magnitude of Pluto was 15. On the other hand, the orbit of Pickering's Planet O was much larger than either Lowell's Planet X or Pluto. Specifically, using the mean distance of the Earth from the Sun as the standard astronomical unit—au—then for Pickering's Planet O, au = 55; for

[25] Pickering was a versatile, innovative and somewhat eccentric astronomer. For a summary of his predictions of undiscovered planets, including extensive references, see William Graves Hoyt in ISIS, Vol 67 (1976), page 551.
[26] See W. H. Pickering in *Popular Astronomy*, Vol. 32 (1924), page 393.
[27] Harlow Shapley (1885-1972) was director of the Harvard College Observatory from 1921 to 1942. His hobby was myrmecology, the study of ants.

Lowell's Planet X, au = 43; and for Pluto, au = 40. So only in a very limited sense was Pickering's prediction more accurate than Lowell's. In fact, the orbit Lowell predicted for his Planet X resembles Pluto's more closely than Le Verrier's prediction for Neptune resembles Neptune's orbit.

Pickering, like Lowell, analyzed the residuals in the positions of Uranus. Unlike Lowell, he also used those of Neptune. Also unlike Lowell, Pickering's analysis was graphical and not analytical.

However, to dismiss the historical importance of W. H. Pickering's prediction of Planet O is a mistake. In 1919, when Pickering's paper on Planet O reached Walter Adams,[28] at that time Assistant Director of the Mount Wilson Observatory; he forwarded it to Harlow Shapley, then on the Mount Wilson staff. Shapley wanted nothing to do with it, and offered it to the janitor. That janitor was none other than 28-year-old Milton Humason[29] who, in addition to his janitorial duties and like Clyde Tombaugh a decade later, had just been assigned to work as a night assistant. Humason saw this assignment, not as an unwelcome additional chore, but as an opportunity for change to a more stimulating occupation. Seth Nicholson,[30] another astronomer at Mount Wilson, noting how close Pickering's planet O was to Lowell's Planet X, suggested to Humason that he could gain astronomical experience by conducting a search for the planet, O or X, with Mt. Wilson's Cooke 10-inch refractor. So in December 1919 Humason began his search continuing it into 1920 with their 60-inch reflector. Humason examined pairs of photographic plates using a blink comparator he had made himself of wood and cardboard, "scraps and scrounge" as he put it. George Ellery Hale[31] lent him the prisms which, in 1919, were hard to obtain. It was impossible to line up the two plates, so Humason just let the images jump. He found nothing.

In 1930, as soon as a preliminary orbit for Pluto was available, it was clear to Nicholson that Pluto's images must be somewhere on Humason's 1919 photographic plates. Later, when the first accurate orbits of Pluto were calculated, it was possible for Nicholson—assisted by twenty-four-

[28] Walter Sydney Adams (1876-1956), stellar spectroscopist at Mount Wilson.
[29] Milton Lasell Humason (1891-1972), noted for his observations of the radial velocities of galaxies, was a high school dropout who became, successively, a mule driver on the trail from Sierra Madre to Mount Wilson, janitor, night assistant, assistant astronomer, then astronomer at the Mount Wilson and Palomar Observatories. In 1950, he was awarded an honorary Ph.D. from the University of Lund, Sweden.
[30] Seth Barnes Nicholson (1891-1963), an enormously versatile and productive astronomer at Mount Wilson, was the discoverer of four of the outer moons of Jupiter.
[31] Hale (1868-1938), solar spectroscopist and founder of the *Astrophysical Journal*, Yerkes Observatory, Mount Wilson Observatory and Palomar Observatory.

year-old Nicholas Mayall[32]—to place the location of Pluto's images on four of Humason's plates within an area about one centimeter in diameter, about the size of a dime. During one week, Nicholson spent one hour per day and all one Saturday morning blinking Humason's plates looking for Pluto's image in that dime-sized area; he found nothing. Mayall spent an entire weekend doing the same; he also found nothing. Then, on Saturday, June 7, 1930, Nicholson at last found Pluto's images. As Nicholson explained to Pickering in a letter of May 11, 1934:

> On the plate of December 27 the shadow of an "S" shaped piece of lint passes between the image of Pluto and that of a nearby star . . . On December 29 Pluto was between two brighter stars and is therefore difficult to see. On December 28 it was even closer to a brighter star.[33]

In short, it would have been practically impossible for Humason, Nicholson, or anybody else to have discovered Pluto on Humason's plates. In 1981, Fred Lawrence Whipple, then in his seventy-fifth year, gave a succinct summary of the situation: "Only the perversity of chance kept the discovery of Pluto from being made by the Mount Wilson astronomers in 1919."[34]

7) *Pluto would not have been discovered in the twentieth century if the Lowell Observatory search had not been initiated.*

This opinion was expressed in April 1930 by William Henry Pickering who wrote:

> All of us . . . owe Lowell a distinct debt of gratitude. Had he not established his observatory at a very considerable expense, and had not his former associates continued his work of searching for the unknown planet, it is highly probable that . . . [Pluto] would not have been discovered for another hundred years.[34]

At this time, Pickering could not have been aware of the nightly activities of Seth Nicholson, the astronomer at Mount Wilson who worked on

[32]Nicholas Ulrich Mayall (1906-1993) was a pioneer spectroscopist of nebulae, globular clusters and galaxies.

[33]From the Mount Wilson archives. We are indebted to Helen Czaplicki for the copy of this letter.

[34]See Whipple's *Orbiting the Sun*, Cambridge, MA, 1981, page 42.
Much of the information in this section is based on Reaves' previously unpublished and informal discussions with Nicholson in 1962, with Shapley in 1964, with Humason in 1968 and with Mayall in 1980. See also the paper by Nicholson and Mayall in the *Astrophysical Journal*, Vol. 73 (1931), page 1.

Pluto with Humason and Mayall. In 1914 Nicholson had discovered the ninth moon of Jupiter, often referred to as Jupiter IX. Since then, Nicholson had kept track of "his" moon with the 100-inch reflector. In a letter of 27 November, 1930, Nicholson told William Hammond Wright at Lick Observatory:

> Had you noticed that Pluto has been in the same field with Jupiter's satellites? The last plate of it [Pluto] showed Jupiter VI not far off. Now I think Pluto will be on the same field with [Jupiter] IX. If Tombaugh had just waited a year we might have saved him a lot of work.

The lesson of history is that—armed with the legacy of confidence—the Lowell team did not wait.

The study of the history of science shows us how science "really" works; the history of science is the study of what people do, not always what they think they did. In the report on the 1931 award of the Jackson-Gwilt Medal of the Royal Astronomical Society to Clyde Tombaugh, the first to identify Pluto's images on the Lowell photographs, he is quoted as saying: "While I was the first to see it, the whole Lowell staff had been working on it for a quarter of a century. I was just lucky, that's all."[35]

Compare this assessment with the comments of William Herschel on his discovery of Uranus in 1781:

> It has generally been supposed that it was a lucky accident that brought this star into my view; this is an evident mistake. In the regular manner I examined every star of the heavens, not only of that magnitude but many far inferior, it was that night its turn to be discovered.

In 1930 it was Pluto's turn to be discovered at Lowell Observatory.

[35] As quoted by the *Coconino Sun* under date of 21 March, 1930, the statement was, "Sure, I was the first to see it, but the whole Lowell staff has been working on it for a quarter of a century. I was just lucky, that's all there was to it."

11

Whither The Sun[1]
1946 – 1994

⎧ CIENCE DURING the war was
⎪ devoted to winning. And, so,
S scientists, including some as-
tronomers, dropped what they were doing, and went to work in the giant
military labs devoted to creating the atomic bomb, jet engines, radar and
military electronics. For example, John Hall, who was later to direct the
Lowell Observatory during its great expansion at the beginning of the
Space Age, worked on radar at MIT's Lincoln Labs.[2]

At the leading American observatories—Lick and Mount Wilson in
California, Yerkes in Wisconsin, Lowell in Arizona, McDonald in Texas
(there was not as yet a big telescope on Palomar, and Mauna Kea wasn't
even a gleam in Gerard Kuiper's eye)—astronomy was temporarily
asleep.[3] Except for the famous story of how Mt. Wilson astronomer Walter
Baade (himself perhaps suspect as a representative, however unwittingly,
of the master race) utilized the temporarily darkened skies above blacked-
out Los Angeles to probe ever deeper into the universe, not much was
happening. These famed institutions, all privately operated, were living
off their investments and biding their time.

In the late forties, the flow of federal funds that was to propel Ameri-

---

[1]This chapter was written by George Wesley Lockwood, a valued staff member of Lowell
Observatory.
[2]While so employed, Hall wrote the book: *Radar Aids to Navigation*, 1947; McGraw-Hill &
Co.; New York.
[3]Kuiper (1905-1973) was born in the Netherlands but became a leading figure of
American astronomy.

189

can science to the Moon and beyond did not yet exist. There was no NASA, only its predecessor, the National Advisory Committee for Aeronautics, which operated a few small aeronautical research centers like Langley in Hampton, Virginia, Lewis in Cleveland, Ames in California, at all of which the studies concentrated on aircraft and immediate military applications, mainly using windtunnels and other laboratory techniques. There was no space science. Except for the German V2 and the rocket studies of Robert Hutchins Goddard,[4] there were no plans or research—most of this came as booty from Germany, captured hardware and brainpower like Wernher von Braun—and when they came, they belonged to the Army at the Redstone Arsenal in Alabama or at White Sands, New Mexico, not NACA. Rockets started as weapons, and after the war, like their creators, they had to be first demobilized and tamed for civilian purposes.

Eventually, of course, there would be civilian money. Enabling legislation for the new National Science Foundation was passed in 1950, and slowly the pace of scientific research began to pick up in the universities and even in the peripheral smaller research institutions like Lowell.

At the end of World War II Lowell was a scientific dwarf. The intense excitement that had brought Lowell into the nation's limelight in 1930 with the discovery of Pluto was definitely over, and its fortuitous discoverer, the photographic technician Tombaugh, was no longer an employee. He was working for the military in New Mexico and would never return. There were only three scientists. The Slipher brothers were prominent, but they were both aging and slowing down. V.M. was still director but after thirty years in the saddle, was largely preoccupied with his private business activities. His younger brother Earl, onetime mayor of Flagstaff, was continuing his careful photographic studies of the planets for which he was to become justly famous. Lampland, quiet, secretive, and obsessively careful to the point of almost complete invisibility, puttered away, leaving ultimately hardly a trace of his activities either in publications or in notoriety.

Meanwhile, the exploration of the universe was being conducted elsewhere. Credit for the recognition of the expanding universe of receding galaxies officially belonged to Slipher, but the fame and glory passed to Hubble for his continuing work on the classification of galaxies. The planets remained the province mainly of the Lowell observers, perhaps because no one else thought the solar system worthy of serious study. In

---

[4] A professor at Clark College in Worcester, MA, he wrote Smithsonian Publication #2450 entitled: *A Method of Reaching Extreme Altitude*, in 1919 at age 37.

astrophysics, Lowell simply wasn't in the game, although this was to change with the arrival of Albert Wilson and Harold Johnson. Planetary science did not exist, per se, though before long an entire division of the American Astronomical Society was to be devoted to it. Kuiper, directing the Yerkes Observatory and observing at McDonald, had discovered the existence of a gaseous methane atmosphere on Saturn's moon, Titan, in 1944, but attracted little notice among mainstream astronomers. Then, as now, fashion ruled the roost, and it was the big telescopes like the new behemoth on Palomar Mountain that made the headlines.

Although no one knew it, wartime technology had brought on the demise of a mainstay of astronomical data recording. The photographic process, by which starlight was captured as blackened grains of silver iodide in a gelatin substrate, was on its way out, although it would not disappear completely for another forty years. Starlight was beginning to be recorded, not by chemistry, but by electronics, indicated first on the galvanometers of the 1920s and then after the war as a wiggly line on a moving scroll of graph paper.

This was difficult technology, but compared with photographic plates it offered 10-100 times the sensitivity and 10 times the accuracy for measuring the brightnesses of point sources like stars. The photocells of the 1920s and 1930s were works of laboratory art—handmade glass envelopes, cathode materials deposited under hard vacuum by techniques that are still mysterious, minuscule currents that could be rendered visible only with the most sensitive galvanometers. There were few practitioners and they were justly famed: Joel Stebbins[5] of the Washburn Observatory, using cells made by his colleague, Jacob Kunz;[6] A. E. Whitford[7] and G. E. Kron[8] at Lick Observatory; and John Hall at the U.S. Naval Observatory. Stebbins advanced the techniques of photoelectric photometry from the primitive and practically useless selenium photocells to the more practical photomultiplier tubes of the postwar era and was the first to measure starlight accurately through colored glass filters.

Astronomical photometry was changed forever, after the war, by the availability of the photomultiplier tube. Unlike the finicky, handmade photocells that produced, at best, a measly current barely detectable with

---

[5]Stebbins (1878-1966), professor of astronomy at the University of Wisconsin, received the 1950 Gold Medal from the Royal Astronomical Society for his work in photoelectric photometry.
[6]Kunz (1874-1939) was a Swiss-born physicist.
[7]Albert Edward Whitford, born in 1905, was a specialist in photoelectric instrumentation.
[8]Gerald Edward Kron, born in 1913, later became director of the U. S. Naval Observatory's Flagstaff station.

the most sensitive equipment, the photomultiplier produced a current a million or more times bigger. The workhorse of the astronomer, the RCA type 1P21 photomultiplier and its generic stepsister, the cheaper 931A, were basically unchanged after 1941 and remained in common use at observatories well into the 1970s. Their applications were at first mainly commercial (motion picture sound decoding) and military (noise sources for radar jamming). Astronomers and physicists used them for photometry and spectroscopy. After the war, thousands were sold in a slightly different format for automatic headlight dimmers for Cadillacs, a commercially exotic idea that never really caught on.

The weak current produced by photomultiplier tubes was nevertheless too small to be utilized directly and, so, required amplification. Astronomers of the day built their own DC amplifiers, a source of additional instrumental grief as these units were every bit as finicky as the phototubes. Doing photometry required knowing the conversion between a given amount of light, as from a star imaged by a telescope, and a corresponding amount of current. The equipment had to be stable with time, and to achieve this was an iffy proposition. The gain calibration had to be verified several times a night, and if anything went wrong, the data were useless.

After amplification, the current was sent to the ubiquitous "Brown Recorder," a clanking monster that produced a jittery inked trace on a scrolling chart of graph paper. (Photons do not arrive in neat rows like marching soldiers, but rather in random clumps more like a street mob. Hence, a noisy record.) This was the one piece of universal commercial hardware that all astronomers utilized; it was robust and reliable but it weighed a ton, being about the size of a big microwave oven. Except for the chronic headache of inkfed systems that hated to work in the cold of observatory domes, they were basically reliable. The usual chart speed was an inch every two minutes, so a night's work produced a chart 25 or more feet long filled with an analog record of jittery squiggles and scribbled notes added by the astronomer to note the time, the star identification, the filters used, the gain settings and all the minutiae of observation.

Now what? From a roll of paper to the *Astrophysical Journal* is a long trip. The term "data reduction" is no misnomer—the night's data literally had to be condensed and translated into numbers, yielding results that filled perhaps only a single page of tabulation. Fifty to a hundred stars was a good night's work. The astronomer, or more typically some low-paid flunky, had to "read" the chart, which meant measuring the heights of the various squiggles on the paper and writing down the numbers.

"High tech" in those days meant measuring the chart with a special ruler calibrated in stellar magnitudes rather than in inches. This saved one laborious step, that of looking up every reading in a table of logarithms, multiplying by 2.5 to produce the astronomer's goofy scale of stellar "magnitudes" that engineers to this day find incomprehensible, applying various corrections for the gain of the amplifier, and so on.

Despite atmospheric and instrumental limitations that limited the precision of photometry to barely 1 percent, and more typically 2 to 3 percent, the detailed computations of data reduction had to be carried out to three and preferably four significant figures throughout. This eliminated the slide rule as an instrument of computation and forced the astronomer to tables of logarithms and mechanical calculators. Hence, the kerchunkety-clunk of the the ten-place Monroematic filled the building from dawn to dusk.

The resulting table of numbers still were far from the final results. These numbers, the so-called "raw" magnitudes still required further processing, all by hand, of course, to account for the peculiarities of the particular equipment used, to correct for the loss of light in the earth's atmosphere (15 percent at the zenith in visible light, twice that much in

Don Thompson inundated by paper from the Monroematic calculator in 1972

the violet). The various sums of products and squares for linear regression were all done with an adding machine, and if a star did not fit, the work had to be done all over again. A night's photometry thus produced a good hard day's work and often more. Today, that task requires only a few minutes to produce a neatly printed summary.

After the war it was clear to the Observatory's sole trustee, Roger Lowell Putnam, that it was high time for new ideas and new blood. Otherwise, the Observatory, which had been quietly dozing for two decades, would simply lapse into a coma as its aging staff withered away.

A 1939 conference held in Flagstaff at the county fairgrounds had caught the attention of the Observatory's director, V. M. Slipher, as well as the trustee, and furnished the genesis of renewed vitality.[9] The focus of the meeting was on agriculture in the plains states but included was a panel on long-range weather forecasting. With civilian aviation gaining momentum, better forecasts were needed.

The chaotic unpredictability of weather had long been a source of frustration for forecasters. If only a Rosetta Stone could be found! One theme that had run like a thread throughout the twentieth century was the possibility, indeed the perceived strong probability, that one key to weather might lie in small changes in the output of the sun. This idea had been vigorously promoted first by S. P. Langley[10] and then by C. G. Abbott, a successor as director of the Smithsonian Institution and its various outposts in North and South America.

Since the turn of the century, Abbott and a small band of dedicated observers had been monitoring the total irradiance of the sun from isolated mountain tops in California, Chile (and for a time from Harquahala Peak in Arizona) using an instrument called a pyrheliometer. These observations had to be adjusted for the amount of radiation lost in the earth's atmosphere, and in particular, for varying amounts of water vapor. The resulting number, the so-called "solar constant" represented the amount of sunlight falling on the earth's atmosphere, expressed, as was the custom in those days, in calories per square centimeter per minute.

It is irresistibly tempting to believe that the solar constant would be

[9]For several years there had been a small amount of U. S. Weather Bureau money available for studying atmospheric circulation on other planets, but Slipher had not cared to look into this matter. The funding would largely have gone to support the work of Adel - a prospect the trustee cared more about than the director.

[10]Samuel Pierpont Langley (1834-1906) was well-known for his experiments in heavier-than-air flight, contemporaneous with those of the Wright brothers. However, he was also director of the Allegheny Observatory and Secretary of the Smithsonian after 1887.

temporarily lowered by the passage of sunspots across the sun's disk, because, after all, sunspots represent regions of lower temperature on the sun. That is why they appear dark. Abbott was convinced that the solar constant fluctuated slightly, and further, that these fluctuations could be associated with specific weather events. The acceptance of sun-weather connections may not have been universal, but the meteorological litera- ture of the 1920s was pervaded by example after example of correlated time series showing various relations between the solar constant, the sunspot number, and various weather and climate phenomena on time scales from days to decades. Abbott was a tireless promoter of anecdotal solar-weather connections and never minded that few were subject to the rigors of statistical tests invented during his tenure at the Smithsonian.

Almost forgotten in the faddish acceptance of solar relations to weath- er is the notorious counterexample of the "Lake Victoria effect," well noted in meteorological folklore. Early in the twentieth century, the respected British climatologist Charles Ernest Pelham Brooks[11] reported a correlation between the level of Lake Victoria in central Africa (a kind of continental rain gauge that is the source of the White Nile) and the fre- quency of sunspots. From 1900 until 1923, the two time series tracked perfectly, leading to the conclusion that solar activity predicted precipita- tion. Unfortunately, after 1923, the two series diverged sharply, never to be joined again.

A recent publication provides yet another sense of deja vu. Two Danish scientists, writing in the journal *Science*,[12] recently presented an almost-per- fect correlation over the past hundred years between the length of the sunspot cycle (that is, the time between successive minima) and global temperature. This result, whether yet another "Lake Victoria effect" or a real relationship, is nevertheless applying heat to the currently fashionable and "politically correct" view that it is man, and man alone, which is responsible for the global temperature rise of the last century. Who knows?

Meanwhile, the idea of solar weather connections was still very much alive, and part of the proceedings of the Flagstaff conference turned to the question of Abbott's measurements and how they might be improved or validated. The astronomers at the conference put forth an idea, not a new one exactly—credit for this dated from the turn of the century—to moni- tor total sunlight by tracking the brightness of reflected light from the outer planets, that is, to study the sun at night!

---

[11] Author of the widely respected, *Climate Through the Ages*.
[12] "Length of the Solar Cycle as an Indicator of Solar Activity closely Associated with Climate" by E. Friis-Christensen and K. Lassen; *Science* 254, p. 698 (1 November, 1991).

The fallout from the 1939 conference proved to be a fruitful turning point for many aspects of the Lowell Observatory. From it came the first-ever government money to arrive at the Lowell Observatory, a program entitled "The Project for the Study of Planetary Atmospheres" funded at first by the U. S. Weather Bureau and later by the Air Force Cambridge Research Laboratories (now at Hanscom Air Force Base, Massachusetts) in the summer of 1949. Much of the work was accomplished in summer sessions by a small team of visiting meteorologists, several of whom have had long and distinguished careers—Ralph Shapiro, a 1943 graduate of Bridgewater College and later project scientist at the Air Force Cambridge Research Center; Seymour Lester Hess, then of the University of Chicago and later Florida State University; Hans Arnold Panofsky, New York University and later Penn State; and Edward Norton Lorenz from MIT. Each of these young participants in Lowell's Planetary Atmospheres program achieved later distinction as professors of meteorology.

The planetary atmospheres project, which entailed studies based mainly on Lowell's photographic collection led to a new observational sub-program that was to endure for forty years with hardly a break that began under the direction of a newly-hired young astronomer, Harold Lester Johnson (1921-1980). Johnson was brilliant, mercurial, impatient and stayed with Lowell Observatory for only seven years, until 1959. His specialty was instrumentation and techniques of photometry, and together with his slightly older contemporary, William Wilson Morgan of the Yerkes Observatory, he devised a system of multi-color photometry, the so-called UBV system. If practical astronomical photoelectric photometry is said to have a father, then that father is surely Harold Johnson, who after two stints at Lowell, worked for a while at the McDonald Observatory in Texas, and went on to the newly-created Lunar and Planetary Laboratory at the University of Arizona in Tucson and then to Mexico.

Actually, the first such observations of Uranus and Neptune were made by Henry Giclas during 1949 using the 1909 40-inch telescope and a primitive photometer. There were lots of problems brought on by an amplifier designed by Johnson that never worked well, and a disconcerting tendency of the telescope to drift away from the star being observed. Nevertheless, Giclas's report, included in the 1950 report to the U.S. Air Force, is a classic presentation of the methods of photoelectric photometry.

Why is it better to monitor the sun indirectly using light reflected from planets, rather than by the more seemingly straightforward direct methods of Langley and Abbott? What was sought was not an actual value of the solar constant—that had already been determined to better than 1 percent—but evidence of small changes of a percent or less. For this purpose, direct measurements were unsuitable owing to the uncertain but necessary corrections that had to be made to account for losses in the Earth's atmosphere. Indeed, a possible interpretation of Abbott's claimed solar-weather connections could well have been changes in the atmospheric transparency, an idea still not completely ruled out.

Observations of the planets neatly skirt this issue, at least to a first approximation, because they can be made differentially with respect to so-called comparison stars lying near the planets' path along the solar ecliptic. When comparing two celestial objects in this way, the amount of light lost in the atmosphere is immaterial, provided that it doesn't change between the two measurements, often only a minute or two apart.

Of course, the planets move with respect to the celestial background, although for Uranus and Neptune, the motions are quite small, only 4 degrees per year for Uranus and half that for Neptune. This meant that each year, new comparison stars had to be selected, and their brightness values tied into a growing network of prior years' stars, slowly creeping along the ecliptic.

The program required careful repeated measurements and a consistent plan of work. The results, prior to Johnson's return in 1952, were not very precise, owing mainly to growing pains in the art of photoelectric photometry. Nevertheless, the first refereed publication on the subject, a study of the brightness variations of Uranus by Henry Giclas appeared in the *Astronomical Journal* in 1954. This paper showed less than a 1 percent variation in the brightness of Uranus since 1950 and less than a 10 percent discrepancy with measurements made in 1927 by Stebbins.

Johnson, perhaps more than coincidentally, used Lowell's 42-inch telescope to redetermine the magnitude of the faintest star that had been observed photoelectrically with the 200-inch telescope on Mount Palomar. He found that it was in error by 0.4 magnitude. It has been a characteristic of Lowell's astronomers to push their instruments and observations to the ultimate limit in accuracy and sensivity. Those with easier access to larger instruments are often complacent in what they do, perhaps because they really do not have to try as hard and feel that others just cannot compete.

Robert Howie Hardie, who together with Giclas published a retrospective analysis of the Lowell photometry of Uranus and Neptune from 1949 to 1954, was only briefly on the staff of the Lowell Observatory. He went on to a distinguished career as a pioneer of photoelectric photometry at the Dyer Observatory at Vanderbilt University.[13] Surprisingly, only these two refereed publications appeared in the standard astronomical literature of the time. Perhaps because of this, the Lowell program attracted little notice, with its most significant findings buried in the far more obscure Lowell Observatory *Bulletin*.

The "solar variations" program continued without interruption through 1966, with a succession of observers and improvements in equipment. In 1961 a Lowell *Bulletin* titled "The Sun as a Variable Star" appeared, summarizing the results of the program from 1953 onward.[14] The author was Krzysztof Serkowski, a young Polish astronomer later to become famous at the University of Arizona for his pioneering work on astronomical polarimetry. The centerpiece and main conclusion of this publication was an eight-year time series of annual mean brightness values for the planets Uranus and Neptune that showed only small fluctuations from year to year.[15]

By observing sun-like comparison stars over this same eight-year interval, however, the Lowell observers had inadvertently produced a data set of great astrophysical interest. Little was known of the intrinsic variability of solar twins. These common dwarf stars lying on the so-called "main sequence" in a diagram of absolute stellar brightness as a function of stellar temperature were, for the most part, thought to be absolutely stable in their light output. Indeed, many were selected in the early 1950s as photometric "standard stars" used to calibrate the colors and magnitudes of other ordinary or variable stars. What Serkowski and Jerzykiewicz (another Polish astronomer recruited into the program) showed was that these ordinary solar dwarf stars, if they varied at all, did not fluctuate by more than a percent or so over eight years.

This program was continued through the thirteenth season, after which it suffered a five-year hiatus until, with some gentle prodding from the meteorologist and paleoclimatologist, John Murray Mitchell, Jr., the Observatory's new director John Hall, reinstated it in 1971, this time assisted by a five-year grant from the National Science Foundation. The

[13]Hardie's 1962 chapter on photoelectric reductions is a classic reference and still in use; see *Stars and Stellar Systems*, Vol. II, U. of Chicago Press. Harold Johnson also contributed to this volume - a chapter on design and construction of photoelectric photometers and amplifiers.

[14]The earlier photometry - 1950-53 - was of inferior quality and was not included in this report.

[15]There were three important contributions to this *Bulletin* series, each entitled; "The Sun as a Variable Star." The first, authored by Johnson appeared in 1959 (LOB # 96). Subsequent papers bearing the same title appeared in 1961 (K. Serkowski, LOB # 106) and in 1966 (M. Jerzykiewicz & K. Serkowski, LOB # 137).

program has continued, with uninterrupted funding from the Division of Atmospheric Sciences until the present time and has produced a unique time series of photoelectric measurements of a number of solar system objects: Saturn's moons Titan and Rhea, Jupiter's moons Io, Europa, and Callisto, Uranus and Neptune.

In an addition, funded for a while again by the Air Force Cambridge Research Laboratories, an eight-year series of precision measurements of several dozen solar type stars produced valuable new data concerning the variability of sunlike stars. This last project again addresses the question of solar variability from a stellar perspective and seems to be showing that the sun's present variability, finally detected decisively from space radiometry, is perhaps abnormally quiescent at the present time.

12

Changing The Guard
1952 – 1968

T HAT "SPIRITED, noble animal,"
Carl Otto Lampland, was born
of Norwegian parents in a log
cabin near Hayfield, Minnesota, on 29 December, 1873, the third of eleven
children. He grew up on the farm, attended country schools, clerked in
the local store, played clarinet in the village band and received his bachelor of science from Valparaiso Normal School in Indiana.

Desirous of further education, he then entered Indiana University[1] and
took up the study of astronomy. After graduating in 1902, he accepted a
job as principal of the high school in Bloomfield, Indiana, but had hardly
learned his way around town when he was offered a position working for
Percival Lowell. The recommendation had come from Miller, Cogshall
and V. M. Slipher, who was by then well beyond the probationary stage of
his employment on Mars Hill. Lampland was prompt to accept the offer
and spent the remaining forty-nine years of his life working at Lowell
Observatory, even earning a mention in the will of his employer.

A letter of 21 December, 1916, is typical of the loyalty which Lampland
constantly showed towards his employer and his observatory. He wrote
to Charles Sargent,[2] director of Harvard's Arnold Arboretum, with whom
Lowell had maintained a close correspondence:

[1] There are actually two such institutions; one at Lafayette is known as Purdue, while that
at Bloomington is known as Indiana University.
[2] Charles Sprague Sargent (1841-1927) was a prolific author on botanical topics, including
the authoritative *Catalog of the Forest Trees of North America*. He was professor of agriculture at Harvard after 1879.

. . . In the death of Dr. Lowell the world of science has lost an able and fearless investigator and leader, but the sense of loss is especially heavy to those who knew him intimately—his many beautiful and lovable traits of character, his whole-souled enthusiasm and wonderful energy in everything he undertook, the lofty example of his life of devotion to science because it was to him a labor of love.[3]

To those who knew him, Lampland was a man of great friendliness and modesty. Indeed, this latter trait was the cause of his seeming scientific anonymity. Yet he was a man of very broad interests, as indicated by even the most cursory look at the slowly acquired but immense personal library which his heirs left at the Observatory after his death. Though intensely loyal to his employer, Lampland obviously had been impacted by the largely adverse reaction to Lowell from his scientific peers.

Soon after his arrival on Mars Hill, Lampland designed an enlarging camera for the 24-inch Clark telescope by means of which he was able to make some of the finest pictures then available, particularly of Mars—a work that brought him a medal from the Royal Photographic Society and an unexpected MA from Indiana University. He then began the first search for the putative "Planet X" that his employer was already convinced orbited out in the distant cold beyond Neptune. This first search was barely begun when its enormity, the extremely narrow field provided by the 24-inch telescope and the lack of sufficient precision in the underlying mathematical analysis made it unprofitable to pursue.

However, the installation of the 40-inch reflector early in the new century was a great advance in available apparatus; it represented an investment Lowell seems to have made principally for Lampland's use, and

. . . the greater part of Lampland's subsequent observing time was devoted to this instrument. It was the first large reflector to be built entirely by the Clark Company; though optically excellent, its mechanical parts and the dome under which it is housed leave much to be desired. For the most rapid work it required the cooperation of three men, but Lampland often used it alone. In addition to the "trans-Neptunian" fields, his thousands of direct photographs include comets, the planets and their satellites, and the principal nebulae of the NGC. He discovered several variable stars, including novae, in spiral and other nebulae; the remarkable nebular appendage of R Aquarii; and changes in the Crab nebula,

[3]In continuing Lowell's interest in botany, Lampland tried importing sequoia seedlings from California to Mars Hill in 1932. They could not withstand the environmental change.

although his plates failed to reveal completely the nature of these changes. His long series of photographs of Nova Persei 1901 show the growth and motion of the nebulosity attached to the star; and he leaves for future study valuable photographic records of variable nebulae, especially NGC 6729 Coronae Australis and NGC 2261 Monocerotis, the latter comprising about a thousand plates taken from 1916 through 1951.[4]

Though much of Lampland's scientific work was unpublished,[5] and thus of lesser value to the rest of humanity, it was not unknown or unrecognized. In 1930, following the discovery of Pluto in which he had been closely involved and the first public announcement of which he had made, the University of Indiana awarded him the doctorate he had not earned in the usual manner. However, his work with thermocouples, in determining the surface temperatures and thermal gradients of the planets—and developing the necessary equipment for this study—was widely known.

A major—but by no means the only—reason for this was that he did it in collaboration with William Weber Coblentz, a close contemporary employed at the National Bureau of Standards in Washington.[6] Coblentz's work was widely recognized; he, too, received the Janssen Medal, one of several prestigious awards. Together, the two scientists devised heat-sensing devices so well focussed and sensitive they could detect the warmth from a burning candle twenty miles away, as well as that from heavenly bodies millions of times more distant.

Prior to the work of Lampland and Coblentz, there was a wide range of speculation about the temperatures to be encountered on distant planetary surfaces. Those deduced for the surface of the full Moon ranged from below the freezing point of water, to above its boiling point. The surfaces of the distant major planets were thought to be nearly incandescent. The work of these two men, however, brought reality into the game and demonstrated that the thermal emission from a distant planet was quite closely related to the intensity of the sunlight it received—logical, but until then unproven and thus unbelieved.

Astronomer William Sinton, a valued associate of Lowell Observatory,

[4]Excerpted from the obituary of Lampland by John Charles Duncan, then of the University of Arizona, published in the Proceedings of the Astronomical Society of the Pacific, 1952; pages 293-6.
[5]Of a total of 940 plates Lampland made of R Monocerotis (also known as Hubble's Variable Nebula) he published only 9.
[6]Coblentz (1873-1962) was chief of the Radiometry Section of the Bureau from 1905 until his retirement in 1945.

in his article on the *History of Infrared Astronomy*[7] notes:

> *Lampland continued to observe the planets and stars beyond the initial data obtained in 1922-24. In fact, between 1924 and 1943 he filled 16 logbooks with infrared observations. But except for abstracts of papers presented at meetings . . . he published nothing. He wrote in his diaries about receiving angry letters from Coblentz about not publishing his data. After Lampland died in 1951 Frank Gifford, a meteorologist working at Lowell Observatory under an Air Force grant, reduced and published Lampland's 1200 Mars observations . . . More recently Arthur Adel did the same for observations which Lampland had made of the daytime comet Skjellerup in 1927 . . . One might consider that the scientific value of most of the remaining observations is now moot. Not so, for the remaining observations are several thousand measurements of Venus on 358 days between 1924 and 1942. The recent discovery of the variability of SO_2 on Venus and the possibility of changes in the cloud cover of that planet make the reduction and publication appear rewarding. My wife and I, aided by a computer, have now completed the reduction.*
>
> *Why did Lampland not publish? Perhaps because his training did not include a Ph.D. But I believe that he loathed taking a stand. He was a Minnesotan of Norwegian extraction. Garrison Keillor in a recent "Prairie Home Companion" show on Public Radio made the point that Minnesotans will say almost anything to be noncommittal. When asked how they are, they respond, "Things could be worse."*

But Lampland's work, published or not, did lead into and provide a foundation for that of Arthur Adel, which took infrared astronomy another giant step forward. More than that, despite his self-effacing and retiring nature, Lampland made Lowell Observatory his life, giving the institution all of his time during his life, and leaving his immense personal library after his death in 1951.

In the early 1950s it became obvious to everyone that a new broom was needed on Mars Hill to rejuvenate the institution. The long-standing staff members suggested a young astronomer on the Mount Wilson staff whom they felt congenial. Thus it was that thirty-five-year-old Albert George Wilson, with a 1947 PhD in mathematics from CalTech came to Lowell from Mount Wilson in June of 1953 "as Assistant Director with the assignment to assist in the initiation of a revitalization program."

[7]Publications of the Astronomical Society of the Pacific, Vol. 98 #600, February, 1986, pages 246-51.

The trustee brought his two older sons to a Planetary Atmosphere Conference in March, 1950. In attendance were: (front row) Roger Putnam, V.M. Slipher, Edson Pettit, Gerard Kuiper, Seymour Hess, and E.C. Slipher; (middle row) Rudolph Penndorf, Arthur Adel, James Edson, Rollin Gillespie, and C.T. Elveg; (top row) William L. Putnam, III, Roger L. Putnam, Jr., unidentified, Henry Giclas, and Carl Lampland.

During the next two years, assuming the directorship a year after his arrival and upon the retirement of V. M. Slipher, Wilson did much to stimulate the premises. Writing to Giclas on 4 April, 1990, almost thirty-six years after leaving Mars Hill, Wilson summed up his recollections:

My first impressions of the observatory were not so much of stagnation as of the remarkable achievements which had been made with minimum resources. There is a definition of a professional that goes, ". . . one who can solve the problem at hand with the means at hand." According to this definition, those at Lowell, right from the founding, were real professionals, much more so than those with fatter budgets and more lucrative connections. I suppose the pioneers had brought from New England their values of hard work along with a strong measure of Yankee ingenuity. Such devices as floating the 24" dome on pontoons attest to this. But the tradition was ongoing. It was decided to aluminize the mir-

ror of the 40" reflector. Henry Giclas' engineering in fabricating the aluminizing tank by adapting local facilities and coming through with first rate results was professionalism at its best. The ratio of accomplishment to budget was high in those days . . .

There was an important opposition of Mars in 1954, best observed from the Southern Hemisphere. We were able to interest the National Geographic Society in a joint NGS-Lowell Observatory "Mars Expedition," which supported E. C. Slipher's photographic research by enabling him to observe the opposition with the large refractor at Pretoria in South Africa. He obtained some important results on the blue clearing[8] and colors of Martian dust storms.

Also during this time period, with the financial assistance of TWA, Lowell Observatory, jointly with Arizona State University, sent Dr. Arthur Adel of ASU to Trincomalee in Sri Lanka to observe the total eclipse of the Sun.

Clyde Tombaugh had calculated that with the 13-inch he could make a survey of cislunar space for the detection of possible natural satellites as small as a tennis ball. The negative results of this program were reassuring to the gestating U.S. space program that damage from collisions with natural meteoroidal material was not a major hazard.

Another cooperative research program during this time frame was with Drs. John Strong and Ralph Sturm of Johns Hopkins University. Sturm had developed a sensitive image orthicon (grandfather of the charge coupled device) and a low noise amplifier which they felt might be used to obtain superior photographs of planets. The results were better than direct photography on a night of average seeing but fell far short of what E. C. Slipher had obtained under optimum conditions.

It had long been recognized that there would be great value in keeping Mars under continuous surveillance for several months before and after an opposition. To this end Lowell Observatory set up the "International Mars Committee" soliciting the cooperation of observatories throughout the world. This organization paid good research dividends in the 1956 opposition.

Other things I recall while serving as director of the Observatory, 1954-1957:

Holding the first ever "Astrobiology Seminar" (with Major Simons and Dr. Strughild, U.S. Air Force);

Cooperating with Disney studios in their filming of movies about the planet Mars and the exploration of Space (Ward Kimble and Bill Bosche);

Setting up the TIAA program for the Observatory staff; Beginning a site sur-

---

[8]Mars really is the *red* planet and when photographed with a blue filter generally gives very poor photographs. However, about 10% of the time, Martian weather permits detail of its albedo to be seen with a blue filter. This condition is not predictable and changes frequently. Generally, the blue clearing seems to bear a relationship to the well-noted Martian dust storms.

E.C. Slipher examines negatives by the north light of the observatory reading room in 1953.

vey to find a location with a darker sky than Mars Hill;
   The gift of the Morgan telescope;
   Observing time exchanges with the Steward Observatory (Ed Carpenter);
   Initiating seminar and guest investigator programs;
   Friday lunches at the Monte Vista Hotel attended by the staffs of the Lowell Observatory, the A.S.U. Observatory, and the Museum of Northern Arizona;
   The death of Mrs. Percival Lowell;
   The death of Mr. Stanley Sykes, pioneer machinist and "professional";
   And many recollections of visiting astronomers, Fred Hoyle, George McVittie, Gerard Kuiper, Bok, Mainel, Popper, . . . It should be remembered that it was in this time interval that Aden Mainel began his site survey for what was to become the Kitt Peak Observatory and Bill Miller of Mt. Wilson Observatory found the glyphs at White Mesa which were interpreted as an Anasazi record of the supernova of 1054.
   In June 1957 in my last week in Flagstaff, the Lowell Observatory came of age in hosting the annual meeting of the Astronomical Society of the Pacific. The revitalization stage was over. Then in October 1957 came Sputnik, followed by the Space Age and a new era for astronomy.

Another man of Mars Hill who bridged the old order to the new was Earl Carl Slipher, eight years younger than V. M. Earl, like his ten siblings, was born on the farm near Mulberry, Indiana, and attended schools at nearby Frankfort. He entered Indiana University in 1902, after V. M. had already started work in Flagstaff. His astronomical career began at the end of his junior year when he was asked to accompany Professor John Miller, chairman of the university's department of astronomy, on an eclipse expedition to Spain in 1906.

In 1905, Percival Lowell, impressed with the two associates he had hired from Indiana University, established the Lawrence[9] Fellowship at Indiana, whereby an astronomy graduate at that institution could intern at his observatory for a year or two and receive a master's degree. John Duncan, later associated with Wellesley College, was the first such fellow in 1905. E. C. Slipher became the second for the years 1906 to 1908. As the 1907 opposition of Mars was another close one where the night time ecliptic was low in the southern sky, Lowell supported an expedition headed by Professor David Todd of Amherst College and E. C. Slipher to observe Mars from the Andes in Peru. They found extremely favorable conditions there and returned to Flagstaff with an excellent record of the planet's seasonal behavior in an extensive series of drawings and pho-

[9]Named after his mother's family.

tographs. This expedition launched E. C.'s career as a lifelong Mars observer. He was granted a M. A. degree as a result of these two years as a Lawrence Fellow and was added to the staff of Lowell Observatory as an astronomer in 1908.

E. C. participated in many more observing expeditions during the ensuing fifty-six years at Lowell Observatory, including four more total solar eclipses in the years 1918, 1923, 1925 and 1932. He headed three Mars observing expeditions to the Southern hemisphere at the Lamont-Hussey Observatory at Bloemfontein, South Africa, in 1939, 1954 and 1956. In 1954 he was instrumental in organizing the International Mars Committee that coordinated observations of Mars from observatories all over the world. This program demonstrated the usefulness of worldwide cooperative surveillance of the planets which led, some ten years later, to the NASA-supported International Planetary Patrol operated by the Planetary Research Center at the Lowell Observatory.

One of E. C.'s first assignments in 1906 was to make plates along the invariable plane[10] with the 5-inch Brashear camera for Lowell's first search for Planet "X." He made over 150 plates before going to South America with Professor Todd. During the next ten years he made many of the Planet "X" search plates with other lenses and telescopes. After the 42-inch reflector was put into operation in 1909, he assisted Lampland in taking search plates with that instrument. But as time went on, E. C. was the only one of the three final associates of Lowell to believe that process of searching to be futile.

Soon after his arrival on Mars Hill, Percival Lowell asked E. C. to help with his planetary observations, not only of Mars, but of Jupiter and Saturn as well. He soon found that E. C. also had exceptionally keen eyesight and the ability to translate what he saw onto paper. Lampland had designed and developed planetary cameras for the 24-inch telescope to take red and blue photographs, for which he had received the gold medal of the Royal Photographical Society in 1909 after taking the first photographs to show the "canals" on Mars. But since Lampland had many other projects going with the reflector, E. C. took over the planetary observing and spent the rest of his career in this field.

In the early 1930s the three senior staff members felt it was time for an update of the Mars program which could best be done by writing a book on the "red planet." It became slow going. Several chapters were written

[10]This is a specifically astronomical term defined as "a central plane of the solar system discovered by LaPlace, passing through its center of gravity at an inclination of 1° 35' to the ecliptic, and independent of the mutual perturbations of the planets."

and then the project evolved into a joke among the wives on the Hill that whenever they could not account for their husbands' observing, they must be working on the Mars book. It did come to pass, however.[11] Beginning in 1960 the space program was planning a fly-by of Mars and this information was sorely needed. In 1960 E. C. headed a U.S. Air Force project at Lowell to update all that was known from ground based observations of Mars. At one time there were eight people working on this project and one of the tasks was to make independent estimates of the blue clearing on over 60,700 images of Mars in order to be able to predict clear Martian skies whenever a probe might arrive. In 1962 his 168-page book *Photographic History of Mars (1905-1961)* was finally published. Two years later his second volume, *A Photographic Study of the Brighter Planets* came out.

E. C. Slipher was appointed the acting director of the Observatory by Roger Putnam in January 1957 upon the resignation of Albert Wilson and served until September of 1958. He was also the de facto liaison between the Observatory and the townspeople of Flagstaff. He was a member of

[11] The staff did not want this project and it was only completed at the insistence of the trustee.

The office of E.C. Slipher, shown here in 1964, was notoriously cluttered with cigar boxes, reports, and negatives which only he could find. It was understandably referred to by observatory secretaries as the "rat's nest."

the Rotary Club and of the City Council and served two terms as mayor of the city.

The younger of the Slipher brothers was an affable man who enjoyed living and was popular in the community. He had many hunting and fishing buddies and participated vigorously in these activities. E. C. was an active member of the Chamber of Commerce for a half-century and was chairman of the "Good Roads Committee" that eventually succeeded in getting Route 117 built between Phoenix and Flagstaff. He served both in the State House of Representatives and the State Senate. From 1935 to 1939 he was a member of the Flagstaff State College Board. Before and during World War II, he was vice-chairman then chairman of the Coconino County Selective Service board.

In June 1919 E. C. married Elizabeth Tidwell from Oklahoma whom he had met when she was the fourth grade teacher at the Arizona Normal School (now NAU). They had two children—Capella, named after the brightest star in the constellation of Auriga, and Earl, Jr. E. C. developed a heart problem some two years before he retired and was forced to slow down in his activities. He died of a heart attack on 7 August 1964, a few months after he retired.

John Scoville Hall (1908-1991) became the fourth director of Lowell Observatory on 1 September, 1958. He was fifty years of age and events would soon show that Roger Putnam had chosen wisely. For the next nineteen years Hall guided the Observatory through a period of unprecedented expansion in staff and facilities. More than any single individual, he molded the Lowell Observatory that enters its second century—clearly the right man in the right place at the right time.

His parents, Nathaniel Conkling Hall and Harriet Rose Lance, lived in Old Lyme, Connecticut.[12] He graduated from Amherst College in 1930 with a degree in physics and promptly enrolled in post-graduate astronomical studies at Yale, specializing in infrared photometry. With his PhD in hand after only three years, Hall held a series of teaching positions—one year at Columbia, four at Swarthmore, four back at Amherst and four at the MIT Radiation Laboratory [see page 189]. Then in 1948 he accepted the position of Director of the Equatorial Division at the U.S. Naval Observatory in Washington. It was this job which led him to Flagstaff and, ultimately, to Lowell Observatory.

Within John Hall's domain at the U.S. Naval Observatory was an

[12]Hall retained his childhood contacts and, after retirement from Lowell Observatory, wrote a paper on certain historical aspects of his birthplace.

unique 40-inch reflecting telescope. It contained the first Ritchey-Chretien set of optics which had been designed to give excellent image quality over a significantly larger field of view than with a conventional Cassegrain telescope. However, this instrument was severely limited by the poor seeing near the nation's capital and was thus rarely used. Hall recognized its value and set out to find a site worthy of the telescope.

In 1952 [see page 169] Hall made the first of a series of site testing visits to Flagstaff. Henry Giclas recalled the nervousness of the courtly eastern-er having to work with portable instruments on dark hilltops while coyotes howled nearby. Coyotes and rattlesnakes were rare in Hall's home turf, but he persevered. Ultimately, a hill five miles west of Flagstaff was selected as home for the U.S. Naval Observatory's Flagstaff station and the 40-inch telescope was installed there in 1955. Arthur Hoag, who worked for Hall in Washington (and would succeed him as director of Lowell in 1978), was named director of the new station but Hall continued to visit Flagstaff periodically to use the telescope in his ever-active research program. During all these visits, extending over several years, Hall became closely acquainted with the Lowell staff—and their trustee.

Throughout his career, John Hall was a productive and pioneering researcher. At Yale he made great strides in developing photoelectric equipment for measuring the brightness of stars. He was the first to cool the photocell with dry ice, thereby lowering its dark current and facilitating the detection of much fainter stars. He was the first to use a photocell to scan stellar spectra, to use a grating to isolate spectral regions for quantitative measurements of stellar colors, and to carry on photoelectric photometry of stars in the infrared. Hall made fundamental contributions to our understanding of the differential absorption of starlight by the interstellar medium. Most importantly, he and Albert Hiltner[13] discovered that the interstellar medium also polarizes the light from distant stars. This fundamental discovery subsequently permitted determination of many properties of the interstellar dust. Hall established a correlation between the amount of absorption and the percentage of polarization—such studies dominated his research to the end.

When Hall arrived to take up the directorship of Lowell late in 1958, he found a smallish institution with somewhat rustic facilities. Almost all the buildings dated from the time of Percival Lowell. Only three modest telescopes—the Clark refractor, "Lampland's" 42-inch reflector and a

---

[13]William Albert Hiltner (1914-1991), a specialist in stellar spectroscopy, was at Yerkes during this period but after 1970 was chairman of the Department of Astronomy at the University of Michigan.

largely homebuilt 21-inch reflector—were available. Even the road up Mars Hill was unpaved. When his wife of twenty-three years arrived on Mars Hill with their two children[14] she is said to have moaned, "John, what have you done?"

The scientific staff at Lowell in those days was equally modest: Earl Slipher was continuing his photographic studies of the planets; Henry Giclas was just beginning his very productive proper motion study; Harold Johnson, an expert in photoelectric photometry (along with William Morgan of Yerkes Observatory), had established the UBV system of stellar photometry still in use today; William Sinton, the newest member of the staff, was continuing the tradition of infrared astronomy at Lowell Observatory pioneered by Lampland and Adel; Braulio Iriarte, who was visiting Lowell from Tonantzintla Observatory in Mexico, worked primarily with Johnson.

Clearly Lowell was in a time of transition—Lampland was dead, V. M. retired and E. C. was winding down. With the funding of his proper motion study by the National Science Foundation Giclas was emerging from the shadow of the older guard. Johnson was a well-recognized stellar astronomer in his prime and Sinton was a talented young researcher out to make his name. Needed was a director with ambition, energy and vision—plus a ready source of funds.

Hall came to Mars Hill at a time when federal funding of research was expanding rapidly. For most of a decade, the Observatory had been receiving minor grants from agencies such as the Weather Bureau, Office of Naval Research, Air Force and National Science Foundation. However, with the launch of Sputnik I on 4 October, 1957, the space race was on and NASA was charged with the task of catching up with the Soviets at all cost. Lowell, the one major observatory in the country which had consistently emphasized the solar system, was ideally situated and Hall seized the opportunity.

As a manager, Hall delegated authority freely for specific projects while retaining a strong grip on the overall course of the institution. Henry Giclas, as the secretary/treasurer, relieved him of many routine responsibilities; but though generous with his own resources, he had been through the long, lean years and guarded the Observatory's assets with vigilance.

Hall's management was inclusive; with monthly meetings of all

[14]John was married to Ruth Carolyn Chandler in 1935. They had two children, Carolyn and Richard, the latter having also become an astronomer, presently associated with Northern Arizona University.

astronomers to discuss projects and concerns, everyone knew what was happening and could be heard. As might be expected for a smiling, friendly man who always had a kind word and ready wit, John Hall was well liked as a boss. He expected the staff to work hard, keep predictable hours and limit coffee breaks both in duration and volume. But he gave the scientists the freedom and the means to pursue their ideas. As a result, their productivity increased tremendously.

Typical of Hall's style was a speech he delivered at a shower for Bill Sinton and his bride, Marjorie Korner, both staff members of Lowell Observatory:

> . . . In order to make the groom as inconspicuous as possible he is expected to wear some sort of funereal-like garb in utter contrast to the finery displayed so conspicuously by his bride. Have you ever heard of a groom wearing his grandfather's wedding costume? . . . We all go to weddings, to admire the bride, her attendants, and the beauty of the ceremony. But would some feeble voice from the balcony cry "Long live the groom?"

. . . Shortly after Robert Millis joined the Lowell staff in 1967, John Hall loaned him one of his photomultiplier tubes. Millis needed this delicate and expensive device to continue his studies of Delta Scuti variable stars. Unfortunately, as the new staff member was aligning the tube for his first night of observation, he accidentally broke it. In the morning, after a number of sleepless hours, Millis went to the director's office and confessed what had happened.

Instead of the expected chewing out, the director simply said; "That's okay. If you never make any mistakes, you aren't doing anything." This typical kindness was greatly appreciated by the recipient and clearly remembered as he helped write this book, some twenty-five years later.

Perhaps Hall, in forgiving the new astronomer, was thinking back to a recent painful incident involving breakage on a grander scale. The 42-inch telescope installed in 1909 originally had what astronomers call a Newtonian optical system. With this arrangement of mirrors, the focus is located on the side of the telescope near the top end of the tube—an inconvenient locale since it required the astronomer to work high in the dome, usually in total darkness.[15] Newtonians are also limited because large, heavy instruments cannot practically be used with these telescopes.

A partial solution to the problem had been implemented fifteen years earlier by Lampland and Clyde Tombaugh. Clyde, a skilled amateur tele-

---

[15] Astronomers have been seriously injured in falling from the Newtonian focus.

The completed 40-inch telescope in 1910. Note the precarious perch of the Newtonian focus.

The Explorers of Mars Hill / 214

The 42-inch mirror after its disastrous cracking in 1964

scope maker, had ground a mirror which sent the focus to the bottom end of the telescope. The primary mirror, however, had no hole in its center, so the light still had to be directed to the side rather than to a more convenient location behind the primary mirror. At best, even the improvement was awkward to use, and soon after his arrival the new director noted the telescope was being used only 20 percent of the time. To rectify the situation, Hall decided to convert the 42-inch to a true Cassegrain telescope, for which a hole would have to be ground through the large, primary mirror.

In late spring of 1964, arrangements were made for Tinsley Laboratories of Berkeley to sandblast a six-inch hole through the center of the mirror, which was about seven inches thick. While a number of optical experts had expressed the opinion that the operation was relatively risk-free, when the cutting was barely an inch into the glass, a distinct "clunk" was heard and the mirror cracked, rendering it worthless.

The calamity did have a happy outcome. Two years later, with funding from the National Science Foundation, Lowell was able to acquire a new 42-inch telescope from Astro Mechanics Corporation of Austin,

John Hall studies the course of construction of the new home for the venerable Perkins telescope on Anderson Mesa in 1960

Texas. This modern and versatile instrument was installed in a new, tall dome at Anderson Mesa and became a major item in the Observatory's research arsenal. Appropriately, like the 40-inch telescope which Hall had rescued from oblivion in Washington, the 42-inch is equipped with Ritchey-Chretien optics, and equally appropriately, it was renamed in 1990 as "The John Scoville Hall Telescope."

As can be seen from the data in the appendices, Hall did far more as director than enhance the physical plant of Lowell Observatory. He rebuilt its staff, brought it revived international respect and, on his retirement in 1977, left it a much stronger institution than he found it. No one said it better than did the man who had hired him, in a letter written midway through Hall's tenure. In 1968, the year after he elected to retire from his part at the helm of the Observatory, Roger Putnam wrote him: ". . . It [Lowell Observatory] has become under your leadership a growing and viable force in astronomy, which it had ceased to be when you first became director."

Ruth Hall was also a great asset to Lowell. A quiet, reserved lady, devoted to her husband and family, she took an interest in the broader Mars Hill family, too. Her part was to revive the social occasions, Christmas parties for the staff children, which had been largely dormant

Arthur Hoag inspects a copy of Bill Hoyt's first book in preparation for assuming the directorship of Lowell Observatory in 1976.

since the death of the founder. It is difficult to imagine a couple better suited to rebuild and represent this old and proud institution than John and Ruth Hall.

The final major change in long-standing custom at Lowell Observatory came towards the end of 1967 when Roger Lowell Putnam (1893-1972) opted to retire as trustee. During his forty-year tenure, he had designated several persons (serially) as his replacement, all in accordance with the will of Percival Lowell (See appendices). His final option was his youngest son, Michael, the family scholar, then thirty-four years of age and a budding authority on classic literature. Roger Putnam had seen the observatory through the long, occasionally halting, recovery from the nearly fatal ten-years after its founder's death.

As noted in Chapter 9, he had stimulated and encouraged great things, almost all of them on a shoestring budget. His final choice of successor trustee was based on his feeling that the working staff on Mars Hill

The old order changed at Lowell Observatory in 1967. Roger Putnam with his successor Michael stands in front of the administration building with retired director V.M. Slipher and director John Hall.

would feel greater compatibility with his scholar son, rather than the businessman or broadcaster. In making the momentous decision to step down, Roger Putnam was stimulated by a minor stroke that had recently incapacitated him for several weeks and from which he was slow in recovering. At age seventy-five, after forty years at the helm of this unique institution and after bringing Michael to Flagstaff for several visits, he turned over the reins completely.

IV

Exploring the
Universe

In 1993, twenty years after its original purpose was concluded, Oscar Saa, the former Cerro Tololo observer for the Planetary Patrol, shows the facility to Kitty Broman. The dome for the Aura 4-meter telescope dominates the skyline where Saa is now facility manager.

13

The Weather On Mars[1]
1968 – ?

B LANK! That's what it was. Absolutely blank. The project people
stared in utter disbelief. Here at
the last moment, something seemed to be seriously wrong with the camera. After a journey of millions of miles, taking over five months,
America's first orbital spacecraft to Mars, Mariner IX, had a stunning
problem. Hundreds of millions of dollars in cost, years of design and fab-
rication, months of navigating the craft to Mars and successful maneuver-
ing into an orbit around the red planet—and now—no pictures. Was the
mission to become a failure at its moment of climax?

No! The problem was with Mars, not with the camera or the spacecraft. The surface of the planet, at that moment, was completely obliterated by a massive dust storm which had started only weeks before. This
simple answer to the project workers' concerns was pointed out by a
planetary astronomer who had been observing Mars with his telescope.
Despite all the careful attention given to the complex technical aspects
of the mission by countless engineers and scientists, no one on the project had thought of following events on the distant objective through a
telescope.

Slowly, over several weeks, the dust settled and the surface features
of Mars reappeared, a few at a time. The planet was beginning to reveal
its secrets, but with agonizing modesty. By the close of the mission, near-

---

[1]This chapter was written by William Edward Brunk, advisor to Lowell Observatory and
staff employee of the National Aeronautics and Space Administration.

ly a year later at the end of October 1972 almost 7000 images of the Martian surface had been received. Mariner IX was justly acclaimed as a great success.[2]

This experience pointed out, once again, the vital importance of ground-based astronomy to the success of the NASA program of planetary exploration. From the start it had been known that it would not have been possible to design any planetary spacecraft without using data on the planets previously obtained from ground-based observations. What had not been anticipated was the critical need for observing activity on the objective planet just prior to and during the mission. Without this information, the tremendous success of the American program of planetary exploration would not have been possible.

It was in just this area, that of planetary monitoring, that the men of Mars Hill made major contributions to NASA's planetary exploration program. Several individuals, almost all names closely associated with Lowell Observatory, played major roles in the critical and ongoing program of planetary photography, the source of much data on the activities at these distant goals.

Changes in the surface appearance of the planets, particularly Mars and Jupiter, had been observed for well over a century. The existence of dust storms, clouds, and other temporal phenomena on Mars had been known among astronomers before the time of Percival Lowell. Many cases of transient phenomena had been noted, but few serious scientific studies had been undertaken. There were many reasons.

Recording details of fine features on planetary surfaces has always been a difficult task. Originally, the only method for recording such features was by means of sketches made by an observer studying the planet through a telescope. This process was as old as Galileo. The accuracy of these sketches was strongly dependant on several variables—the skill, experience and artistic ability of the observer, the quality of his telescope, and most importantly, the clarity and stability of the atmosphere when and where the observations were made.

The introduction of photography into astronomy at the end of the nineteenth century made it possible to obtain accurate, permanent records of planetary features. However, such photography was hampered by two major problems. The image of the planet on the photographic plate was so small that only the major surface features were distinguishable. Also, the chemical emulsions used were slow, requiring

[2]Mariners IV, VI and VII had all been fly-by efforts as far as Mars was concerned; they continued on into interplanetary space.

very long exposure times. The Earth's atmosphere, which acts like a dirty window in constant motion, blurred almost all the finer detail. Though faster emulsions came along and later electronic devices improved quality a hundredfold, difficulties with the age-old problem of "seeing" still plague planetary photography to a great extent.

Photography of the planets was initiated at Lowell Observatory in the early 1900s. In 1903 Carl Lampland, who had joined the Mars Hill staff the year before, made many successful photographs of the planets. E. C. Slipher, who arrived in 1906, carried out an ongoing program of planetary photography from 1907 almost until his death in 1964. He had the unique experience of photographing Mars during four favorable oppositions— 1909, 1924, 1939 and 1956. Mars is always a very small object when viewed from Earth. It appears largest in angular diameter when viewed near opposition, when the planet is directly opposite the Sun in the sky as seen from Earth. Oppositions of Mars occur approximately every 26 months, but since Mars has a more elliptical orbit than Earth around the Sun, the apparent diameter of the planet varies from one opposition to the next, being as small as 1/140th the diameter of our Moon at the poorest to 1/75th the lunar diameter at the most favorable oppositions.

With the increased use of photography in astronomy, the planets were photographed frequently but discontinuously from observatories around the world. But, dedicated planetary photography, except in the case of Mars, had received little concentrated effort.

When planetary exploration became a real possibility with the development of space flight, it became apparent that there was a serious gap in our understanding of the planets. This gap occurred as a result of the shift in scientific interest by professional astronomers from planets to stars, which took place in the early 1900s. The new, large telescopes which were being constructed provided significant increases in the ability to gather light, of great benefit for observations of the more distant stars, but provided little or no gain in capability for making better planetary observations. To fill the gap in planetary knowledge, serious attempts were made, starting in the late 1950s, to study the planets using astronomical techniques that had been developed for the study of the stars. Interest was also revived in observing surface features on the planets by means of photography.

Planetary photographs were difficult to obtain on many of the large telescopes. Observing time on them was very limited as these facilities were fully scheduled for stellar observations and many astronomers considered planetary studies as second rate science. Also, the sizes of the planetary images obtained with these telescopes were small and the fine details of the images obtained were too small and distorted by both diffraction of light in the telescopes themselves and motions in the atmosphere across the aperture of the telescope. However, considerable effort was directed by a few professional astronomers towards increasing the quality of planetary photographs and the amount of information obtainable from them.

Starting in 1960, Dutch-born Gerard Peter Kuiper (1905-1973) of the Lunar and Planetary Laboratory at the University of Arizona, one of the few professional astronomers with a strong interest in planetary studies, designed and built a 61-inch telescope specifically optimized for planetary photography. He spent considerable effort in picking a site for his telescope, optimizing its design, improving the quality of its mirror and controlling air currents and temperatures inside the dome to reduce adverse seeing effects on the quality of the planetary images.

In 1961, a dozen years after he left Flagstaff, Clyde Tombaugh was professor at New Mexico State University but still extremely active in planetary imaging. He worked with a 24-inch reflecting telescope designed for and dedicated to high resolution planetary photography. An unique feature of this small telescope was the relatively large images of the planets it delivered without using a complicated and light-absorbing optical system. The telescope had an effective focal length of f/75, which produced relatively large images of the planets. Most telescopes dedicated to non-planetary work have ratios of f/15 or smaller. The physical size of a planetary image with the f/75 telescope is five times larger than that obtained with a f/15 instrument. With Tombaugh's new telescope, an image of Mars at a favorable opposition was 5.6 millimeters in diameter, compared to 1.1 millimeters with a 24-inch f/15 telescope.

Tombaugh, with his dedicated telescope, undertook a regular photographic patrol of the brighter planets observable from his location. Kuiper, on the other hand, interested in as high resolution images as possible, took high quality and frequent, but not continuous, photographs of the brighter planets.

In spite of these efforts, there were still very large gaps in the photographic coverage of any planet. The reasons were many; lack of interest in planetary astronomy by most professional astronomers, lack of telescopes capable of high resolution, large scale planetary images, difficulty in obtaining significant amounts of observing time on existing large telescopes, and the problem of having good weather and a steady atmosphere at an observatory over a period of days or weeks. Even under ideal conditions, observations with one telescope were only possible for a period of six to eight hours a night, with the result that the entire surface of any planet could not be observed from a single site because the rotation period of the planet was longer than this time interval.

A concentrated study was undertaken to solve some of these problems for Mars, during the favorable opposition of 1954. In 1953 an International Mars Committee was established to bring together astronomers who were equipped to make contributions to the Martian observational program and to coordinate their cooperative efforts. E. C. Slipher and A. G. Wilson of Lowell were joint secretaries of the committee. A major achievement of the program was the International Photographic Planetary Patrol. Seven observatories located at widely separated longitudes participated in the patrol with supplementary photography from three additional stations. Approximately 40,000 exposures were taken of Mars.

While these efforts were generally successful, the results were not quantitatively easy to interpret as the program depended completely on voluntary participation using whatever telescopes, cameras and filters were available. However, the program was sufficiently successful that it was repeated for the even more favorable opposition of 1956. At that time E. C. Slipher served as committee chairman.

The criticality of understanding the atmospheric dynamics of the planets prior to any planetary mission was recognized early in the space program. The only reliable method for studying planetary atmospheric dynamics was by observing the motions and changes of surface features on the planets from photographic images of the planets.[3] By 1958, when NASA was established, thousands of such images existed but they were scattered throughout the world, mostly at the observatories where the images had been obtained. This material had to be collected and assembled before it could be analyzed To accomplish this, Dr. Kuiper formulated the following resolution which he presented to Commission 16, *Physical Studies of Planets*, of the International Astronomical Union (IAU) at their August 1961 meeting in Berkeley, California;

The committee appointed by Commission 16 on "International Collaboration for Planetary Observations" desires to facilitate international collaboration on planetary studies by the eventual establishment of at least two data centers, one in the United States and one in Europe; and meanwhile requests observatories

[3]It was with this very goal that the U. S. Weather Bureau had become interested in working with the astronomers on Mars Hill back in 1939.

having large collections of planetary photographs to make these available for such studies as require a full coverage in longitude.

This resolution, passed by the IAU General Assembly, also authorized the President of Commission 16 to take the necessary steps to cause its implementation. The Lowell Observatory at Flagstaff, Arizona, and the Astrophysical Section of the Observatoire de Paris at Meudon, France, were selected as the respective data centers and officially requested to establish repositories for their hemispheres. Lowell was chosen because of its long specialization in planetary work and since it already possessed a large fraction of all existing planetary photographs, a collection estimated by Dr. Slipher in 1962 at some 30,000 to 40,000 plates.

Armed with the IAU resolution, Dr. John Hall, Director of Lowell Observatory, approached NASA in 1964 with a proposal to establish and operate a Western Hemisphere Planetary Research Center at the Lowell Observatory in accordance with the IAU resolution. Recognizing the need for the study of planetary surfaces to support the program of planetary exploration, NASA provided the necessary funding for the construction of an "adequate and attractive" building to house the collection as well as providing the operating funds for the first three years of activity.

In early 1965 Dr. Hall wrote to the directors of all observatories in the United States having significant libraries of planetary images. He requested permission to examine their collections and make arrangements to obtain copies for the Planetary Research Center. If satisfactory to the observatory directors, arrangements were made to ship their planetary plate collections to Mars Hill, where they were copied and then returned. A print of each photograph thus obtained was then sent to the planetary center at Meudon, which in turn sent the Lowell Center copies of all images they obtained from Eastern Hemisphere observatories.

On 1 October, 1965 forty-one year-old Dr. William Alvin Baum, formerly of the Mount Wilson and Palomar Observatories, was appointed Scientific Director of the new Planetary Research Center. After arriving at Lowell Observatory, Baum initiated research programs on such planetary problems as the time variability of the Martian Polar Caps, the growth and motion of Martian clouds, etc, using the full planetary image collection. However, a major problem was soon encountered in attempting to use this collection for research. The images, having been taken at various observatories with different telescopes and cameras, had a wide variety of image sizes. Useful quantitative study of planetary dynamics using this material required that all images be of the same size. Baum solved this problem by developing a special planet image projector built so that individual images could, by projection, be precisely matched in position, size and rotation. A coordinate grid could also be placed in the plane of the viewing screen to permit the precise measurement of the location of features on the planet's surface.

It quickly became apparent that, although many problems could be approached using the existing historical collection, any serious attempt to study planetary atmospheric dynamics would require a dedicated, multi-observatory, observational program using telescopes with similar instrumentation and covering all longitudes. With the idea of just such a program, an international planetary patrol, Baum approached NASA. He suggested that Lowell Observatory establish an international network of telescopes to carry out a continuous photographic monitor of Mars and the brighter planets. The idea was to obtain 'round the clock, high quality photographic coverage of each planet with identical image sizes. Whenever possible, existing reflecting telescopes with approximately 24-inch mirrors would be used.

All sites would be furnished with custom cameras to provide identical scale images. At longitudes where observations were needed and no existing telescope was available, new and dedicated instruments of 24-inch aperture would be installed. These would be of the same optical configuration as that used by Tombaugh at New Mexico State University. To further ensure uniformity of images, all photographs would be made with the same type of film and identical filters. All film would be furnished to each site by Lowell Observatory, exposed at the site, and returned, undeveloped, to Lowell where it would be processed, copied and filed. A finished copy would then be returned to the observatory where it was taken. The patrol photography would be carried out independently at each site under the direction of the observatory director and using local astronomers as observers.

The International Planetary Patrol became operational in 1969 with the following five well-known observing stations participating: Mars Hill in Arizona, the Magdalena Peak Observatory of New Mexico State University, Mauna Kea in Hawaii, Mount Stromlo Observatory near Canberra, Australia, the Republic Observatory in Johannesburg, South Africa, and Cerro Tololo in Chile. In 1970 the Astrophysical Observatory at Kodaikanal in southern India, joined the network and was followed the next year by Perth Observatory in Western Australia. Magdalena Peak participated only for the first year of the Patrol and the Mt. Stromlo Station was deactivated after Perth came on line. Three of the Patrol sites,

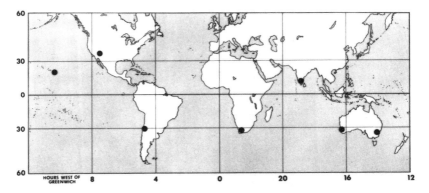

Planetary Patrol stations surrounded the world in 1970.

The Planetary Patrol station at Perth occupies a unique elevated dome shown here nearing completion in 1971.

The Explorers of Mars Hill / 228

Mauna Kea, Cerro Tololo and Perth—were equipped with brand new 24-inch Boller & Chivens telescopes. New optics, identical to those in the Boller & Chivens instruments were installed in existing telescopes at Lowell and Kodaikanal.[4]

Initial emphasis in the patrol program was on Mars with Jupiter being observed whenever possible. Venus was observed from some of the stations in the near ultraviolet where the featureless planet showed darker, cloud related markings. Saturn was also observed at some stations, but not until later in the program. During the first three years a total of 56,000 image sequences were obtained, with each sequence containing fourteen images. An essentially full patrol was continued until 1976 and reduced versions, consisting of observations from one or more sites, through 1990.

Data from the planetary patrol were used in different ways. The patrol images, along with the historical data from the Lowell files, were used to study variable phenomena on Mars, Jupiter and, to a limited extent, Venus, quite independent of any spacecraft missions. The patrol images were also used to directly support the NASA missions to Mars and Jupiter.

The discovery that Mariner IX had arrived at Mars just over a month and a half after a major dust storm obliterated the martian landscape alerted the Viking project team that they needed to better understand the weather on Mars in support of the Viking mission. A rush program was initiated to determine, based on past records, what the weather conditions there could be expected to be at the time of the Viking I encounter with that planet. Patrol images of Mars were used in conjunction with historical images from the Lowell files to predict the Martian weather at the arrival of the Viking spacecraft. Careful study of all the available data led to the conclusion that the Martian weather should be good at encounter for both Viking I and Viking II.

During the actual missions, continuing status reports were supplied to the project office and scientists. As Viking I neared Mars, patrol results were being provided to the project on almost a daily basis to assure complete knowledge of conditions at the time of orbital insertion and lander deployment. When the Viking spacecraft finally arrived, conditions on the planet were ideal. But this was no real surprise; astronomers on Earth had done their forecasting well and assured the Viking project that such would be the case.

Later, during NASA's missions to Jupiter, the planetary patrol results

[4]The results obtained varied from quite good, where the staff was highly motivated, such as Perth, to rather dismal at Kodaikanal, where bureaucratic difficulties made a shambles of the effort.

were again utilized. In this instance, concern was not on the occurrence of dust storms; rather the appearance and disappearance of features in the planet's upper atmosphere. The striking differences between spacecraft images of Jupiter taken during the Pioneer encounters of 1973-4 and those during the Voyager encounters of 1979 surprised many scientists, but not those planetary astronomers who had been observing significant changes in the general appearance of Jupiter during that time period.

At Lowell Observatory, routine photographic monitoring of the planets has given way to a more diversified program of focussed research. However, in the case of Mars—and for obvious emotional and historic reasons—Lowell astronomers continue to record the behavior of the planet at each apparition.

14

Chasing Celestial Shadows[1]

ASTRONOMY, by and large, is a passive science. The astronomer is an on-looker who must view the objects of his study from afar. Physicists, chemists, biologists and even most geologists can poke, prod, dissect and otherwise probe their subject matter with a direct, hands-on approach. The astronomer cannot perform such experiments and must arrive at his conclusions by deduction from distant observations.

Fortunately, nature sometimes performs powerful experiments for us. Lowell Observatory astronomers, in keeping with its founder's heritage in solar system astronomy, have experienced great successes in capitalizing on certain of these random opportunities—particularly those celestial events called occultations.

Every object in the solar system casts a shadow in the light of every star in the sky. These may be so faint as to be unobservable to the naked eye, but they are nonetheless very real shadows. These dark beacons sweep through Space and, on occasion, one will pass across the surface of the Earth. An observer in the path of that shadow, who happens to be watching the two converging objects, will, as that shadow sweeps across him, see the star disappear for a time behind the occulting body and then

---

[1]This chapter was written by Dr. Robert Lowell Millis and Lawrence Wasserman. Millis, a native of Illinois who received his PhD from the University of Wisconsin in 1968. He has been employed at Lowell Observatory ever since, serving as its director since 1990. Wasserman received his PhD from Cornell in 1973 and has been on the staff of Lowell Observatory since 1974.

pop into view again on the other side. By carefully monitoring the brightness of the star at each instant throughout the occultation, it is possible to determine properties of the intervening solar system object that are otherwise unobtainable from the surface of the Earth. It is also possible, as in the case of the more prominent events called solar eclipses, to learn things about the more distant star.

William Baum (then at Mt. Wilson Observatory) and Arthur Code[2] were the first to successfully exploit the occultation technique when they photoelectrically recorded the 1952 occultation of Sigma Areitis by Jupiter. Baum and Code, based on their observations of this event, calculated the temperature of Jupiter's upper atmosphere. While Baum did not seriously continue occultation observations after coming to Mars Hill, this process was seized upon by Robert Millis, who was hired as a graduate student, to work with Baum. Millis, trained as a photometrist, was interested in time-variable phenomena; when he went to the telescope, he wanted to see things happen.

After completing his thesis on short-period variable stars Millis collaborated with Otto Franz[3] and Donald Thompson[4] in measuring the rotational brightness of outer planet satellites. These variations resulted from differences in reflectivity and color of the material on the various faces of the satellites and thus provided clues as to the texture of their surfaces. In 1970 Franz and Millis went on to study the brightness behavior of the Jovian satellite, Io, as it came out of eclipse by Jupiter. The satellite had been reported to be anomalously bright for a few minutes immediately after emerging from Jupiter's shadow—presumably due to the formation of frost on Io's surface while in eclipse. Franz and Millis were unable to find evidence of this post-eclipse brightening and brashly stated in print that they believed it did not exist, thereby setting off a controversy that went on for more than a decade.[5]

In 1973-74 Millis helped coordinate a world-wide campaign to observe another type of occultation phenomenon involving Io and the other three

[2]Code received his PhD from the University of Chicago in 1950 and has been professor of Astronomy at the University of Wisconsin.
[3]Franz, Austrian by birth and education, has been on the staff of Lowell Observatory since 1965 and has been the Secretary/Treasurer of the Observatory since 1985.
[4]Thompson graduated from the University of Southern California in 1969 and has been a valued and versatile member of the Lowell Observatory staff ever since.
[5]As this book is in preparation, current occultation studies of Io, by a younger member of the Lowell staff, Marc Buie, are confirming a much different explanation for this occasional brightening - volcanic action - a function of the enormous internal stresses of Io due to strong and varying gravitational attractions between Jupiter and its other satellites.

large satellites of Jupiter. During those years the orbital plane of these four—essentially coplanar—satellites swept across the Sun and the Earth. With the orbits all edge-on, pairs of satellites frequently fell along a line passing through the Earth and causing an occultation as one satellite passed through the shadow of another, producing a form of eclipse. Millis and his collaborators used telescopes at Lowell, plus the planetary patrol telescope at Perth Observatory,[6] to record dozens of these so-called mutual events. These observations, when combined with those from other stations, yielded significant improvement in mankind's knowledge of the satellites orbital parameters.

At about this time, Wasserman joined the Lowell staff. He had just completed a thesis on the interpretation of observations of yet another occultation of a bright star by Jupiter—the 1971 event involving Beta Scorpii. With the arrival of Wasserman in 1974 and that the year before of a young Paris-trained British astronomer, Edward Bowell,[7] the Lowell occultation program really blossomed. Millis concentrated on overall strategy, instrumentation and expedition planning. Wasserman developed computer programs for predicting occultations and for analysis of the resulting data. Franz provided experience in precision astrometry—an essential skill for establishing the exact location on Earth from which a particular occultation would be observable. Bowell developed an unique data base, which he continually updated and refined, of the orbital parameters of all known asteroids. And all four, plus other Lowell astronomers and assistants, spent much of the next twenty years chasing occultations.

These expeditions have invariably produced suspense, sometimes great adventure, and on several occasions, noteworthy scientific discoveries. Besides all corners of North America, the occultation teams have chased shadows in Ecuador, Australia and the Canary Islands. In 1977 one of these ventures led to the discovery of previously unsuspected rings around the planet Uranus; in 1984 another trip enabled the Lowell team to make a precise determination of the size and shape of the largest asteroid, Ceres; and—most rewardingly for Lowell Observatory—in 1988 the occultation team was able to prove beyond doubt that Pluto has an atmosphere.

One can learn much from these events. For small objects (asteroids,

---

[6]Lowell Observatory has maintained a close and cordial relationship with the Perth Observatory since the first introduction of the Planetary Patrol.
[7]Bowell received his PhD from the University of Paris in 1973 and has been on the Lowell staff ever since.

Occultation crew of Wasserman, Millis, and Franz prepare to go on the road in 1977.

small planets and planetary satellites), one can accurately measure the sizes of these bodies by simply determining the time that the star disappears behind the object and the time that it reappears at different locations on the Earth. For larger bodies which have atmospheres, the light from the star fades slowly as it is bent by the atmosphere.[8] Lastly, rings can be detected and their absorption measured by the way they absorb the light of the star.

How are these events predicted, and actually observed? Practically speaking one cannot, for example, simply watch any given star until some nearby celestial body passes in front of it. Rather, one must be able to predict these events well in advance—and with a high degree of accuracy—in order to be able to observe them. Given a machine-readable catalog of stars (typically 300,000, or so) and an ephemeris for the planets, satellites, and asteroids for which we want to predict occulations, interested observers can run time ahead in the computer and see which

---

[8]The rate of such bending is a measure of the temperature structure of the atmosphere on the occulting body.

object's shadows will pass over which part of the Earth. Surprisingly, this does not take too long; it can be done for 200 objects for one year in a few hours of computing time.

Since the star which casts the shadow is effectively infinitely far away, what passes across the Earth is an exact silhouette of the planet or asteroid which casts it. For a large planet like Jupiter or Saturn with diameters of tens of thousands of kilometers, the shadow has a diameter many times the size of Earth. Thus, when the shadow passes across us, anyone on the side of this planet which faces that one will see the occultation (assuming that they are also at a location where the sun has set). On the other hand, a typical asteroid has a diameter of only a few hundred kilometers, and a shadow much smaller than Earth's. When such a shadow passes across our planet, it will describe a relatively narrow track on the ground. Only those observers who are within that narrow track, and also where the sun has set, will be able to observe the occultation.

Even then, things are not so simple. The catalog of star positions contains errors and uncertainties. The models which predict the future locations of planets, satellites and asteroids also contain errors and uncertainties. The result is that we do not know exactly where the shadow will fall on Earth's surface. The uncertainty varies, but is typically measured in hundreds of kilometers. This is not very critical for an object like Jupiter, where the shadow is several times the size of Earth, but when the shadow itself is only a few hundred kilometers (or less) wide, the uncertainty may be as big or bigger than the shadow width itself. The only way to reduce the uncertainty is to take photographs which contain both the object and the star it will occult at some time before the actual event. Accurate astrometric measurements of the separation of the star and object on the plate can then provide a correction to the ephemeris of the asteroid and catalog position of the star. This procedure will usually reduce the uncertainty to less than half of a typical asteroid's shadow track width. But the procedure requires that the object and the star both be on the same plate. Due to the relatively fast motion of the asteroid in the sky this condition is often not met until only a week or two before the actual event.

A second problem, which is also particularly severe for the smaller shadows, is that its track may not pass over any existing telescopes. This means that observers have to pack up and go to where the event will occur. For this reason, Lowell Observatory has acquired three portable telescopes and data recording systems which can be easily transported. The systems use small motor generators for electrical power, so that they can be set up at any location.

There are two more-or-less typical scenarios. The first, for occulations by large bodies and the second for small bodies. For a large body the early prediction is usually reliable enough to make firm plans; the event will probably be visible everywhere on Earth that the object is above the horizon in a dark sky. Thus, one can plan to take portable telescopes to well-defined locations, or apply for time on fixed telescopes. Last minute astrometry is not usually required and the only remaining problem is the local weather.

On the other hand, for a small body, observers normally have just an approximate initial ground track location. A go, no-go decision can be made only after plates are taken and measured a week or so before the event so that the observers can decide if the track is accessible. At that time, weather statistics help to decide where, along that track, to try observing the event. If possible, more plates are obtained to try to further refine the location of the track. When all calculations are completed, a few days before the event, the portable systems are loaded in trucks and teams of observers head off towards the general area where they hope to set up their gear and watch. This location can change as the weather predictions change. If all goes well, the observing systems are placed at locations staggered across the track and under clear skies.

A few case studies indicate that observing such events is often far from straightforward.

In 1973, Gordon Taylor, of the Royal Greenwich Observatory, predicted that Uranus would occult a star on 10 March, 1977, and that the shadow, four times larger than Earth, would cover our entire planet. The event was predicted to be visible from all of the land area surrounding the Indian Ocean, including eastern Africa, India, southeast Asia, Japan and the western half of Australia. Since this would provide an opportunity both to measure the size of Uranus and to determine the temperature structure of its atmosphere, plans were made at several observatories to observe this event. From Flagstaff, Millis requested time on the 24-inch Lowell Observatory Planetary Patrol Telescope in Perth, Australia. At MIT, a group led by James Elliot[9] obtained time on the Kuiper Airborne Observatory (KAO).[10] Observatories in Japan and India also planned to study the event.

[9]Elliot, a professor of Astronomy at MIT, received his PhD from Harvard in 1972.
[10]The KAO is a converted C141 Cargo plane fitted with a 36-inch telescope. Its primary use is to carry infrared instruments above the absorbing water vapor of the Earth's lower atmosphere, but on special occasions it is used to make unique observations above any possible cloudiness in the visible wavelengths.

As it happened, Uranus passed close to the star in late January, before reversing its apparent motion in the sky and occulting the star in March. Plates were obtained by the U. S. Naval Observatory Flagstaff Station at the time of this close passage and measured by Wasserman and Franz of Lowell Observatory. The new measurements indicated that the shadow of Uranus would pass considerably farther south than the original prediction and as a result would not cover the entire Earth. The northernmost part of the shadow would only pass over the southernmost part of the Earth—an area which included only southeastern Africa, the southern Indian Ocean and extreme southwestern Australia. The uncertainties in the prediction were large enough that it was no longer positive that the event would occur at Perth. It was not even certain that the KAO which had planned to chase the event over the Indian Ocean from a base in Perth would be able to fly far enough south to observe it either. After much discussion of the risks involved, it was decided that the benefit of observing this event was worth the risk and both Millis and the MIT group decided to try.

At Perth, Millis started observing early in case the prediction was in error in time as well as position (an error in the position of Uranus along its direction of motion would change the time of the event) A half hour before the predicted time of the event, the signal abruptly dropped by 30 percent and then recovered eight seconds later. This could not be the planetary occultation, nor could it be a cloud since the sky was clear. What had happened? During the next quarter hour, there were four more sudden dips, shallower and shorter than the first. Observations continued, but nothing else was seen until the sun rose. The actual planetary occultation had not occurred at Perth.

Meanwhile, on the KAO, well out over the southern Indian Ocean, the MIT team was far enough south that they were (barely) within the planet's shadow. They saw a similar set of five short dips, then the expected gradual dimming as they passed into the shadow, then a gradual brightening half an hour later as they came out of the shadow and then another set of five short dips. Both Millis and Elliot initially thought that they had seen a "swarm" of satellites around Uranus which had generated the unexpected dips. It was only a few days later, after Elliot had returned home, that he noticed that the small dips before the main occultation lined up with the dips that appeared after the main occultation. Thus, the dips had to be due to continuous rings and not individual satellites. Final confirmation came when the locations of the three of the MIT events were matched with three of the events observed by Millis as well as with the events seen by two other observers and all found to coincide. Later, it was noted that Millis had not seen two of the MIT events because he was centering the image of Uranus in his photometer during the time that they occurred. The two other dips seen by Millis were found to correspond to very shallow events in the MIT data which were missed in their initial examination of the data. Thus, seven rings around Uranus were discovered by this occultation.

The 13 November, 1984, occultation of a star by the largest asteroid, Ceres, was discovered at Lowell by searching photographic plates for stars ahead of the motion of the asteroid which might be occulted. This task was attempted because no potential events were found via the standard catalog search. This process is not performed for every possible asteroid because it is far more time consuming than the computer search.

Ceres has a diameter of about 950 km (580 miles)—far smaller than Earth. The predicted track, crossing the Caribbean, Florida, southern Texas and Mexico was estimated to be uncertain by a half to a full track width. Preliminary plans were made to observe the event from western Mexico using portable telescopes. The final locations of the sites would be determined after last-minute astrometry and with one eye on the weather. A collaborative effort was set up with the University of Arizona to convoy into Mexico. Lowell Observatory provided four observing teams, each with their own vehicle, telescope and data system. The University of Arizona provided two more observing systems and teams, each with their own vehicle and also sent an observer to a fixed telescope in Mexico. The Lowell effort included seven astronomers and staff members plus two astronomers from the University of Maryland and one from the University of Central Michigan. The University of Arizona provided five astronomers. For this event, last minute astrometry was provided by Arnold Klemola[11] of Lick Observatory. His result indicated that the actual ground track would be somewhat south of the original predicted track.

The expedition got off to a very shaky start. Although everything had been cleared with the Mexican government to import the necessary equipment temporarily there was a hang up at the border. As a result, the combined teams were allowed to enter Mexico, but were promptly put into a holding pen for most of a day. This area had obviously been used to hold drug smugglers as there were several vehicles there which had been torn apart looking for drugs as well as a jail which could have come out of an old west movie set. Finally allowed to proceed, they were now a

[11]Klemola, a long-time staff member at Lick Observatory, received his PhD from the University of California (Berkeley) in 1962.

day behind schedule. The six vehicles went in convoy to Culiacan where they began to split up, eventually ending up in six locations along the Mexican coast, from Culiacan in the north, to Chamela in the south. The original plan had been to set up all of the stations somewhat further to the south but due to the delay at the border there was not time. As a result, it was thought that all six stations would fall to the north of or within the northern part of the shadow of Ceres. However, for reasons still not understood, the last-minute prediction was incorrect, and the actual track was further north than the prediction indicated. Thus, by good luck, which had not seemed so at the start, the six portable stations were very well distributed across the shadow zone.

Not everything went perfectly, however—it never does on these things. One of the observing teams chose to set up near a microwave tower about two miles off the main highway halfway between Culiacan and Mazatlan. After the event, at about midnight, their vehicle would not start due to a dead battery. They slept in the car and in the morning hiked back to the highway and hitched a ride with a local truck driver back to their hotel in Mazatlan. When the other teams returned to convoy home, they were able to go back to the vehicle and jump start it. One of the teams had problems with mosquitoes and were also clouded out; one had problems with ants; and one team tried to drive onto a Mexican naval base but found themselves looking down gun barrels until they were able to explain what they were doing—then they were finally permitted to observe.

Despite all these difficulties results were obtained from this occultation. Coupled with additional observations in Florida and the Caribbean, the astronomers were able to reduce the uncertainty in the diameter of Ceres from over 10 percent to about half of a percent.

The 9 June, 1988, occultation of a star by Pluto was originally predicted by Douglas Mink[12] and Arnold Klemola in 1984—again by searching ahead of Pluto's orbit on photographic plates. An accurate prediction for Pluto is extremely difficult due to its small size and great distance. Their prediction was rough, and only indicated that an occultation might happen on the side of the Earth roughly centered on Australia. Later plates by Klemola indicated that the event was likely to occur in the southern hemisphere but were still very uncertain. Predicting the event was further complicated by Pluto's satellite, Charon, which is about half the size of its parent. The motion of Charon introduces a small but definite wob-

---

[12]Mink received his advanced degree from MIT in 1974 and is on the staff of the Smithsonian Astrophysical Observatory as a scientific applications programmer.

V.M. Slipher and Stanley Sykes hang a wreath at the Lowell mausoleum on March 13, 1955, the centennial anniversary of Percival Lowell's birth.

ble into the motion of Pluto, which further confuses the astrometry.

By early 1988 a number of newer astronomic measurements by Klemola convinced the Lowell team that the occultation track would cross the south Pacific Ocean and part of Australia, but it was still uncertain by at least its own diameter, or about 2000km (1200 miles). Nevertheless, Millis started planning to observe the event using a portable system somewhere in Australia and the MIT group (again led by Elliot) made plans to observe somewhere over the south Pacific with the KAO.

The only way to pin down the actual track was to do as much last minute astrometry as possible, preferably using the U.S. Naval Observatory's 61-inch astrographic telescope which is capable of providing extremely accurate plates. However, this telescope has a very small field and the required astrometry must be done using a "bootstrap" process with a secondary net of stars which was provided by Klemola. Fortunately, the weather cooperated and the U.S. Naval Observatory was able to obtain plates on every night but two from May 26th through the event. The resulting astrometry pinpointed the occultation to within about a third of a diameter, allowing both Millis and the KAO to place themselves within the track.[13]

As with the Uranus occultation, there was a surprise. The Lowell astronomers had expected to measure the size of Pluto in the same way they had measured the sizes of asteroids. However, the drop at the occultation was slow, rather than abrupt, indicating a tenuous, but real, atmosphere on Pluto. In addition, the drop consisted of two distinct sections, indicating a layering of some sort in that atmosphere. It is not yet certain if the layering is due to a haze (or cloudiness) in the atmosphere, or to a change in the temperature structure below a certain level. But, of one thing Lowell Observatory scientists were certain—once again they had learned something new about Percival Lowell's planet.

[13]MIT also led a major effort to predict the path of this event using a novel method of scanning strips in the sky with a CCD camera. While there may yet be promise in this concept, unfortunately it did not provide any useful data on this occasion.

V

All Work and
No Play

Stream crossings such as this one on the Little Colorado in 1915 were often too much for horseless carriages.

15

Mars Hill Apocrypha
1894 – ?

MANY TALES abound in the local folklore of the last century in Northern Arizona concerning the patrician Percival Lowell, his associates, and their larger-than-life activities in the world. Some of these tales are true and they all make good reading.

Shortly after Lowell's seven-passenger Stevens-Duryea arrived at Flagstaff in 1911, Dr. Lowell, Judge Doe, E. C. Slipher and Stanley Sykes made a trip to the Horseshoe Cienaga in the White Mountains. There were no intercity highways in those days, so the automobile, complete with camping gear and provisions, was loaded on a flatcar and taken to Holbrook, where the trip began.

Sykes said[1] that on these trips Judge Doe would never sleep in a tent, but commandeered the extra wide back seat of the car and, wrapping himself in a bearskin, sat up with a bottle of whiskey nestled between his feet for ease of access in case of cold. There was never room for all the gear inside the car, so boxes, tents and bags were strapped to the running boards and onto the folded-back top of the touring car. The banging pots and flapping bread sacks created quite a sight as the big, red car sped through the small communities south of Holbrook.

Somewhere near the Mormon farming community of Snowflake, their route crossed an irrigation ditch. On the way south, the ditch was crossed

[1] To Henry Giclas, whose recollections are a delight to hear.

245

C.O. Lampland, Judge E.M. Doe, and V.M. Slipher at a picnic site in 1912

without incident. However, on the way back, two days later, about a mile before reaching the ditch, the road turned into a bog and the expedition's vehicle got stuck in the mud. They finally found a local resident with a team of horses to pull them out and assist the car around a detour. When they got back to the irrigation ditch they found that the 3 x 36-inch tires of the automobile had cut through the sides of the ditch sufficiently that water had been running down the road since they had first passed.

Percival Lowell always maintained that he had exceptionally good eyesight.[2] It was on this trip that both Stanley and E. C. Slipher concurred with this claim. Lowell described details of distant houses, churches and windmills long before they became discernible to the others.

The astronomer was also an amateur botanist who enjoyed collecting specimens and would often invite Stanley to accompany him on such trips. One favorite collection spot was Sycamore Canyon, southwest of Flagstaff. Both here and in Oak Creek Canyon Lowell found a previously undescribed species of ash tree that was later named for him by Professor

[2]He had spotted "clumps" in the rings of Saturn, which Viking photographs identified - eighty years later - as additional but "embedded" Saturnian moons.

Sargent.[3] On these trips Stanley would often make sketches of the scenery and transform them into oil paintings, some of which still hang in Lowell Observatory offices.

Stanley Sykes, for whom all the trustees of Lowell Observatory gained an enormous respect, also had a well-developed, even malicious, sense of humor. When he and his brother were not flooded with work at their "make and mend shop"—the usual condition—they would take advantage of visiting newcomers, as Giclas remembered:

On occasion, since Stanley and Godfrey looked very much alike, they would get up a two-bit or fifty-cent bet that one of them could beat some visitor around the block on a bike. The racers would start off together, but then the brother on wheels would purposely fall behind, letting the challenged feel he had it made. The other brother, in the meantime, would ride through the alley behind their shop and, as the newcomer rode by, pop out behind him with a fresh burst of speed and win the race. In the meantime, the first brother had returned to the shop and would compliment the winner on his great ability to overtake such a fast-pedalling opponent.

In those days, the wooden sidewalk in front of the shop was about two feet above the ground. Another prank was to drop a coin through the cracks in the boards and then tell one of the newer folk of losing some money under the sidewalk. Invariably someone would crawl under the walkway to retrieve the cash, at which moment one of the brothers would find it necessary to dump a pail of water out the door.

The mending business was never too rushing, especially during the winter, when the whole town seemed to hibernate. This condition allowed the Sykes brothers to explore elsewhere in Arizona. Godfrey had heard so much about the Gulf of California he wanted to see it for himself, and thus built himself a vessel on their lot and then hauled it by rail to Mellen, where the railroad crossed the Colorado River to California. With an ex-sailor friend, Godfrey then drifted to Yuma and on into the gulf. That was the winter when Stanley, with his friends, had gone placer mining on the Hassayampa.[4]

Tragedy almost befell them both. Godfrey's boat, Hilda, *burned to the waterline about 140 miles south of the river's mouth and the sailors had to walk north, almost dying of thirst before reaching the Hardy River, some 170 miles from the accident scene. Upon arrival back at Yuma, Godfrey learned in a newspaper account of Stanley's death in the flood [See Chapter 7].*

[3]The *Fraxinus Lowellii* forms a 20-foot tree with deeply furrowed bark and 3-inch leaves.
[4]This passage is from the unpublished Giclas reminiscences.

I was, of course, greatly distressed at the news; but I was quickly relieved by receiving a letter from him in which he stated that although his drowning had undoubtedly taken place, the condition was only temporary, and that he was drying out satisfactorily and was helping to bury other victims of the flood who had been less fortunate.[5]

The brothers, equally undaunted, were back in Flagstaff by springtime.

Ben Doney[6] was a veteran of the Civil War who ran a lot of cattle in the Flagstaff area—somewhat in the manner the Sykes brothers had once done; he was also a sometime city councilor. To him any fence was an insult, if not worse, and he was mightily offended by that which Percival Lowell had caused to be placed around his premises on Mars Hill. On 16 January, 1908, the *Coconino Sun* reported a plan for his revenge under the heading: "COUNCIL PROCEEDINGS."

It was moved by Doney seconded by Manning and carried that the following resolution be adopted: It is hereby resolved by the mayor and common council of the town of Flagstaff, that the town clerk notify R. J. Kidd, probate judge and ex-officio trustee of Flagstaff townsite, to commence the necessary proceedings forthwith in the proper courts to eject one Percival Lowell and his employees and agents from the seventy-five (75) acres of townsite land, now fenced in by the west part of the south half of section sixteen, of said townsite and claimed by him as his own property.

It is further resolved that the said Kidd be notified, that unless he takes necessary proceedings as hereinbefore required the said town of Flagstaff will commence the necessary proceedings in the courts to compel him to be present at the council meeting to be held January 9th, at 1 o'clock p.m.

There were other problems on the town council's collective mind at the meeting of 31 December, 1907, thus reported: the absence of Mayor Leo Verkamp from certain Council meetings, the resignation of Councillor James Harrington, and, the necessity to wreak a vengeance on the Babbitt Brothers comparable to that voted for Lowell Observatory. However, the same issue of the *Sun* carried a response by the mayor, who could obviously give as well as receive:

[5] *A Westerly Trend*, page 224.
[6] After his death in 1932, a park area of yellow pine timber, northeast of Flagstaff, was named in his memory.

In the first instance please allow me to cite that I feel myself more or less of a martyr, for having the patience to read the disgraceful communication. The diction, I am sure emanated from the foul mind of some scavenger of waste baskets whose brain extends no farther back than the roots of his tongue. It bears the ear marks of having been written by one who should rely more on his bodily strength, or rid the community of his person. This illiterate no doubt roams the streets in the guise of a friend to all, or as it is sometimes termed "Carries water on both shoulders," and his aim is to incite trouble by his dirty and filthy work . . .

I am opposed to taking from Mr. Lowell, rather, attempting to take, any lands given him by the previous council. The Lowell observatory is an ornament and big advertisement to the town, and any city would offer big inducement to have him with them. Mr. Lowell does a volume of business in the community, and the advertisement given Flagstaff would warrant any courtesies extended him.

The mayor went on to defend the Babbitt Brothers who had "at all times upheld Flagstaff . . . are one of the biggest tax payers and have done much to further the interests of the town . . ." District Judge Richard Sloan soon heard both cases and found little merit in either, decisions that were greeted with what the *Sun* stated to be ". . . the general approval of the public regardless of any technicalities . . ."

After her husband's death the departures of Mrs. Lowell back to Boston were an invariable hassle because she had so much to take with her. Besides her voluminous luggage, which she would not allow to be checked, there were extra shopping bags of salvaged wrapping paper from the butcher shop, string and paper bags that she wanted to use in Boston. She carried scraps of food left from her last few meals and a bowl of something she wanted to have on the way. She would use a chair car from Flagstaff but moved to a berth for the night, after the train left Albuquerque, as the Pullman charges were considerably less that way—and then she moved back to the chair car in the morning. Staff members could never understand her need to be so frugal, except for their awareness that she took care of "several extravagant nieces and nephews."

The principal delight of the old house, now generally unoccupied, and one that lasted over many subsequent years, concerned the contents of the amply stocked wine cellar that Percival Lowell had accumulated.[7] Cases of the finest domestic and imported wines of all types, cordials,

[7] A detailed inventory of its original contents can be determined from the numerous invoices submitted by H. Jevne Co., Importers, of Los Angeles, that are preserved in the observatory archives.

Constance Lowell at the 24-inch telescope in 1918

Constance Lowell tending the Lowell ash tree near the completed mausoleum in 1925

cognacs and demijohns of table wines had also been supplied by M. E. Bellows' Sons of New York. Whiskey came in three-gallon oak kegs.

Shortly after the repeal of Prohibition in the early thirties, Mrs. Lowell wanted to take some of her preferred wines and cordials back to Boston. The laws then were very strict about shipping alcoholic beverages across state lines and the only way it could be done was through one licensed wholesale dealer to another. Mrs. Lowell found a friendly wholesaler in Boston who agreed to receive a shipment for her, and Henry Giclas[8] asked Harry Moore, a wholesaler in Flagstaff to ship it. Mrs. Lowell then selected three cases of what she liked best and they were shipped off, after which she gave the remaining contents of the wine cellar to Dr. Lampland to dispose.

Since the V. M. Sliphers were teetotalers and Mrs. Lampland was an active Womens' Christian Temperance worker, Henry and a few others became the beneficiaries of most of the wine that remained after a thief had broken in under the house to abscond with the whiskey.

[8]From whose delightful, but unpublished, *Reminiscences* many of the better parts of this book have been derived.

The loss of the whiskey was an inside job. One Schaal,[9] an occasional carpenter and maintenance man on Mars Hill, who also worked for V. M. on some of his rental properties downtown, had learned of the wine cellar. During his sporadic work sessions around the Observatory he took time to cut an opening into the insulated walls of the cellar and would help himself to some of the whiskey during each visit. Obviously he was not a connoisseur of wine because he left it behind, but before the theft was discovered all the oak kegs had disappeared. Thereafter, Dr. Lampland doled out what was left sparingly. It was always a delight; about noon on festive days like Thanksgiving, Christmas or New Year's, he would come by the staff residences with a bottle of wine from the cellar, apologetically saying he did not know if it was a good one or not. But it always was.

Left in the cellar till last was a case of half-bottles of Chateau Royale from the Cresta Blanca winery of Napa Valley, California. It had aged in the bottle for over forty years and turned out to be the most banner wine of all. Henry shared the final two bottles with the other wine connoisseurs of Flagstaff, Joe Babbitt, Norm Sharber and John Stilley, who all agreed it was the finest they had ever tasted.

One of the biweekly rituals of life on Mars Hill in the earlier years was to start the water pump on the way up, and shut it off on the way down from work. There was never enough city pressure to deliver water to the top of the hill, so it was necessary to pump to the twelve thousand gallon masonry tank on the Hill. The pumphouse was about one-third of the way up. In early January 1912 there was no snow and the temperature had not been above -17°F for over a week. All the water pipes, even three feet underground, froze, and it was necessary to build fires over the pipeline all the way up the hill.

Shortly after this a new pumphouse was built in Thorpe Park below the hill and a three-phase electric piston pump installed. This was much improved and even had a lightning-proof circuit breaker that shut the pump off if one phase of the power went out. This also made it possible to turn the pump off from the main switch at the top of the hill and saved a lot of overflowing if Stanley Sykes or his assistant were not ready to go home when the tank was filled.

However, two or three times a year, the cylinders of the pump had to be repacked. One of the fun things in the 1930s was to decide how the

[9]Giclas thought his first name was George, but in all records, even his own letters, he is referred to only by his last name. The initial appears once.

Old pump shelter on the slope of Mars Hill, 1933

pump was to be repacked. The mundane way was to use a homemade piano wire corkscrew and pull the packing out of the glands bit by bit. The more spectacular way was to "blow the packing." To do this, the valves on the supply line and on the uphill side were both closed and the glands holding the packing down on each cylinder were unscrewed. Then the three hundred feet of water in the pipe leading uphill (ca 150 lbs./sq. inch) was released by opening the uphill valve. There was always much argument as to who would perform this final task, as that person could never hope to escape the pumphouse before receiving a super shower, but the process finished the job in a hurry.

Margaret Edmondson, daughter of the distinguished sidereal astronomer, Henry Norris Russell, recalled visiting Lowell Observatory at age thirteen, where . . .

Dr. Lampland was my favorite of all the staff. He always had time to talk to me and tell about the places and people thereabouts . . . One of his stories concerned a time when he was observing at the 42-inch, by himself as usual. The dome was then of canvas, so one could hear any noises outside. This night he

Snow accumulation can be severe on Mars Hill in midwinter as this 1932 photograph indicates.

heard an unusual noise going round and round the dome. He was too busy with his work to go outside to investigate, but when morning twilight put an end to his observations he left the dome to find the tracks of a pair of mountain lions that had been prowling around the dome all night.[10]

At that time there were no paved roads in northern Arizona, and the single track roads were dusty and bumpy. Once or twice each summer, generally during a period when the moon was full and the seeing poor, there would be a major expedition of the whole observatory staff along with visiting friends and family to some more distant scenic area. In 1927, with the Russell family participating, the trip was north to the junction of Grand Canyon and the canyon of the Little Colorado, which was a very isolated area with only sheepherder's tracks for access.

From there, it was back to the "highway"

[10]Quotations in this segment are from a letter by Margaret Edmondson to the author, under date of 2 February, 1992, some fifty-eight years after she had married Frank Kelley Edmondson, another of the Lawrence fellows who played such an important part in the scientific history of Lowell Observatory.

. . . where we gassed up at the Gap trading post - our five cars almost using up the filling station supply - and then on north to Lee's[11] Ferry, which at that time was still a ferry. Mrs. E. C. Slipher's younger sister, Inez, was visiting from Oklahoma, and she happily rode the bonnet of the observatory's Model T, like a bucking bronco.

By the time they reached the ferry, it was dusk and they found the vessel was operated only during daylight hours, so the party had to camp again, and with limited food.

In the morning light we all followed the steep and dusty dugway down to the river where the ferry crew took us, car by car, across. This ferry was the only means of crossing the Colorado river for hundreds of miles. It was a flat boat with a plank top which was hand-pulled across by ropes at an angle, facing upstream into the brawling, red-waved current. Once on the north side we went down to the rancher who had guest accommodations. The rancher's wife produced a magnificent breakfast for everyone, and when Dr. Slipher asked her how much he owed her for us all, she replied, "I don't know; I never fed an army before. Would five dollars be all right?

In this same vein the second trustee of Lowell Observatory once recounted to his children a minor detail of an automobile trip he took across the country in 1915, following his graduation from college, when hard paved roads were a national rarity. The route chosen was planned to take him to the West Coast by way of his Uncle Percival's observatory. On his way west, hungry, dusty and tired from repeated engine repairs and patchings of tires, he stopped for a meal at Holbrook, a few hours east of Flagstaff. The emporium of his choice enticed him in with a window sign that read:

> MEALS—25 CENTS
>
> GOOD SQUARE MEALS—50 CENTS
>
> REG'LAR GORGE—1 DOLLAR

[11] John Doyle Lee, a companion of Brigham Young, carried his first passengers across the river here in January 1872.

Stanley's nephew, Glenton, told about a trip Lowell undertook to the Grand Canyon in pre-Stevens-Duryea days. According to his account,[12] around the turn of the century, this great natural wonder was only casually known. Prospectors, hunters, some sheep ranchers and a few hardy sightseers who got through on buckboards were about all the visitors the great gorge had.

It seems that Percival Lowell, for whatever reason, decided to make a trip up to the South Rim. He secured the services of a guide one summer, a man believed to be part Basque and part Indian, often referred to in the early West as a half-breed. Lowell wanted to make good time, so he decided to make the trip in a special rig drawn by two Arabian horses. It is reported that several people advised him against this procedure, saying that the stock was unused to the terrain and they might have trouble. Lowell, however, proceeded as planned.

It was getting late in the fall and they ran into foul weather, a cold early winter rain. The damp weather persisted for several days until the country was completely saturated and their conveyance was virtually bogged down in the clay type mud of the juniper woodlands. Despite the poor going, they had gotten pretty well along and were only some twenty miles from the south rim of the Grand Canyon. However, conditions

[12]This vignette is paraphrased from Glenton G. Sykes' unpublished typescript of 1957, entitled; *Scraps from the Past.*

Far left: This circa 1914 work train at the Flagstaff station still has an old paint job on the tender.

Left: Percival Lowell, the futurist, tries Arizona's first skis in 1910.

were getting worse, they were running short on horse feed and early one morning Lowell decided to call the expedition off. He sent the man out of their tent to get the stock, but neither guide nor horses ever returned. Lowell waited around camp for several hours, decided he had been "stood up" and made plans to walk out of the country.

He knew pretty well where he was and, being an astronomer, had little difficulty in figuring his direction, so he headed for Williams, the closest point of civilization and some forty miles to the south. The weather stayed mostly bad and he had quite a trip, floundering through miles of mud and uncharted juniper wilderness. His course, however, was good and he finally arrived at the frontier logging town on the railroad. It was quite a feat of endurance and, as Godfrey[13] told him later, should have been recorded as one of his greater accomplishments.

His troubles were not yet over; he had obtained his initial objective, Williams, but was entirely without funds, for his "guide" had not only absconded with the team, but had also taken his employer's spare cash. Lowell had hit Williams on a holiday weekend and every lumberjack, sheepherder and railroader for some distance around was in town to celebrate.

It was difficult to get anyone's attention long enough to discuss serious business, especially like lending money. The aristocratic New England

[13]Glenton's father.

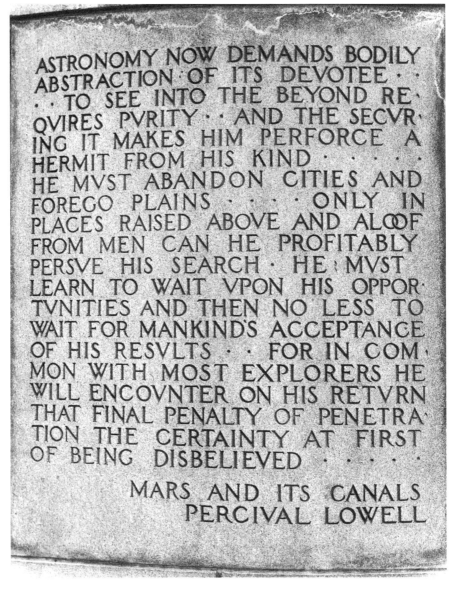

ASTRONOMY NOW DEMANDS BODILY
ABSTRACTION OF ITS DEVOTEE · ·
· · TO SEE INTO THE BEYOND RE·
QVIRES PVRITY · · AND THE SECVR·
ING IT MAKES HIM PERFORCE A
HERMIT FROM HIS KIND · · · · ·
HE MVST ABANDON CITIES AND
FOREGO PLAINS · · · · ONLY IN
PLACES RAISED ABOVE AND ALOOF
FROM MEN CAN HE PROFITABLY
PERSVE HIS SEARCH · HE MVST
LEARN TO WAIT VPON HIS OPPOR·
TVNITIES AND THEN NO LESS TO
WAIT FOR MANKIND'S ACCEPTANCE
OF HIS RESVLTS · · FOR IN COM·
MON WITH MOST EXPLORERS HE
WILL ENCOVNTER ON HIS RETVRN
THAT FINAL PENALTY OF PENETRA·
TION THE CERTAINTY AT FIRST
OF BEING DISBELIEVED · · · · ·

MARS AND ITS CANALS
PERCIVAL LOWELL

Inscription on Lowell mausoleum

The Explorers of Mars Hill / 258

scientist, unaccustomed to this inconvenience, finally lost patience and strode into the largest saloon stating that he was Percival Lowell of Flagstaff and demanding that the establishment advance him a loan. Covered with mud, his clothes in rags and unshaven for a week, he looked like anything but Percival Lowell. The barkeep did not believe it, thought he was just another celebrant, and great merriment was had by all. One fellow, however, said he would take a chance and flipped Lowell a twenty-dollar gold piece, which it was reported was repaid double the next day.

With money in hand, Lowell went over to the depot and wired Flagstaff for a special train - at that time he owned a substantial block of stock in the Santa Fe RR. Eventually a locomotive arrived pulling a lone passenger car, Lowell stepped aboard in all his muddy and solitary glory and, to the cheers of most of the saloon patrons who had gathered at the depot to wish him Godspeed, was transported home to Flagstaff in state.

"The balance of the net income shall be used for carrying on the study of astronomy, especially the study of our solar system and its evolution, at my observatory at Flagstaff, Arizona and at such other places as may from time to time be convenient . . ."
From the will of Percival Lowell, February 21, 1913

The Explorers of Mars Hill / 260

Appendices

Partial New World Genealogy of Percival Lowell
(and pertinent relatives)

GENERATION	NAME	SPOUSE
1.	Percival Lowle (1571-1664) emigrated from Bristol, England, in 1639; settled in Newbury, Massachusetts	Rebecca ? (d. 1643)
2.	John Lowell (1595-1647) emigrated with his father - farmer in Newbury	Mary ? (d. 1639)
3.	John Lowell (1629-1694) cooper - moved to Boston	(3) Naomi Sylvester (1649-1695)
4.	Ebenezer Lowell (1675-1741) cordwainer - resided in Boston	Elizabeth Shailer (1680-1761)
5.	Rev. John Lowell (1704-1767) pastor - 3rd Church of Newbury	Sarah Champney (1704-1756)

6.	John Lowell (1743-1802) [Old Judge] federal judge - Boston	(1) Sarah Higginson (1744-1772)
7A.	John Lowell, Jr (1769-1840) [The Rebel] philanthropist & writer - Roxbury	Rebecca Amory (1771-1842)
8A.	John Amory Lowell (1798-1881) financier & traveler - Boston	(2) Elizabeth Cabot Putnam (1807-1881)
9A.	Augustus Lowell (1830-1901) financier & philanthropist - Boston	Katharine Bigelow Lawrence (1832-1895)
10A.	Percival Lowell (1855-1916) writer & astronomer - Flagstaff Trustee - 1894-1916	Constance Savage Keith (1863-1954)
10A.	Abbott Lawrence Lowell (1856-1943) lawyer & educator - Cambridge	10B. Anna Parker Lowell (1856-1930)
10A.	Katharine Lowell (1858-1925)	(1) Alfred Roosevelt (- d. 1891) (2) James Bowlker (- d. 1917)
10A.	Elizabeth Lowell (1862-1935)	10C. William Lowell Putnam (1861-1924) lawyer and businessman - Boston
10A.	Amy Lowell (1874-1925) poet	

6.	John Lowell (1743-1802) [Old Judge] federal judge - Boston	(2) Susanna Cabot (d. 1777)
7B.	Francis Cabot Lowell (1775-1817) industrialist - Boston	Hannah Jackson (d. 1815)
8B.	Francis C. Lowell, Jr. (1803-1874) industrialist & humanitarian - Boston	Mary Lowell Gardner (1802-1854)

263 / *Appendices*

Staff Members of Lowell Observatory

Those who have labored on Mars Hill in the employ of Percival Lowell from the founding of his observatory through its first century.[1] Herewith are included only those of the scientific and supporting staff who remained for two years of continuous fulltime employment or a reasonable alternative. Not included are the names of the dozens of transient scholars or summer assistants who have graced the Lowell Observatory premises over the years.

NAME	TENURE	DUTIES
ADEL, Arthur	1936-1942	Astronomer
BAILEY, J. W.	1925-1928	Maintenance
BARCUS, Loretta Anne	1974-1977	Computer programmer
BARNUM, Edward Fred	1992-----	Equipment specialist
BAUM, William Alvin	1965-1991	Astronomer, P R C Director
BEISER, Antoinette Sansone	1991-----	Librarian
BENNETT, Arthur	1926-1928	Astronomer
BERTRAM, Raymond Cole	1978-1982	Research assistant
BLECHA, Robert	1958-1963	Instrument maker
BOOTHROYD, Samuel Latimer	1898-1899	Research assistant
BOWELL, Edward Leonard George	1973-----	Astronomer
BOYCE, Peter Bradford	1963-1973	Astronomer
BROOKS, Bivian Lou	1965-1968	Janitor
BROWN, Constance	1929-1930	Secretary
BRUMBAUGH, Dorothy Lee	1979-1991	Librarian
BUCKINGHAM, William Leslie	1991-----	Public program director
BUCKWALD, Kurt Alan	1991-----	Public program
BUIE, Marc William	1991-----	Astronomer
BURNHAM, Robert	1958-1979	Proper Motion Study assistant
BUS, Schelte John	1984-----	Observer
BUS, Ellen Susanna	1984-1987	Observer
BUSBY, Daniel M.	1969-1973	Engineer\designer
CAPEN, Charles Franklin	1953-1983	Public program
CARRIGAN, William Thomas	1905-1909	Computer - Planet X
CHASTAIN, James	1967-1980	Electronic technician
CLARK, Joyce Beatrice	1963-1965	Secretary
COGSHALL, Wilbur Adelman	1896-1897	Astronomer
COOK, Eugene H.	1966-1989	Maintenance
COOK, Jay R.	1953-1967	Maintenance/construction
COOK, Nona O.	1966-1974	Darkroom assistant
CREE, Allan	1930-1932	Secretary
CROWFOOT, Norman C	1975-1979	Computer programmer
CULP, Harold	1967-1977	Instrument maker
CURTIS, Frank C.	1909-1916	Instrument maker/Research asst.
DALES, Neil George	1967-1969	Computer assistant
DARWIN, James Lanier	1988-----	Instrument maker

1. Due to the considerable expansion of staff that has accompanied new construction and temporary projects after 1992, this compilation actually represents only 98 years.

DeGROFF, Kent Gilbert	1964-1968	Assistant astronomer
DeVAUCOULEURS, Gerard Henri	1957-1958	Astronomer
DOUGLASS, Andrew Ellicott	1894-1901	Astronomer (in charge)
DOWNUM, Philip Benson	1959-1961	Computer assistant
DREW, Daniel Abbott	1896-1898	Astronomer
DRYER, Ivan M.	1961-1963	Darkroom assistant
DUNCAN, John Charles	1905-1906, 1912	Astronomer (Fellow)
EDEN, Laurene	1926-1929	Secretary
EDMONDSON, Frank Kelley	1933-1934	Astronomer (Fellow)
EDWARDS, Earl A.	1914-1916	Research asst./Instrument maker
EVANS, Mary Lou Kantz	1972-----	Assistant/archivist
FAURE, Bruce Q.	1965-1971	ACIC Observer
FERGUSON, Holly Merki	1972-1977	Darkroom assistant
FISCHBACKER, Gordon E.	1966-1969	P R C
FRANZ, Otto Gustav	1965-----	Astronomer/Secretary-treasurer
FREDRICKS, Lawrence William	1959-1963	Astronomer
FRIEND, Kathleen Anne	1979-1981	Fund raiser
GADDIS, Sonja Deason	1955-1960	Secretary
GALLAGHER, John Sill	1986-1989	Astronomer, Director
GARDINER, Alan J.	1954-1956	Electronic technician
GICLAS, Henry Lee	1931-----	Astronomer, Executive officer
GIDEON, James Albert	1964-1969	Instrument maker
GILL, Thomas B.	1914-1916	Computer
GREGORY, Sharon Ann	1964-1966	Secretary ACIC
GROGAN, Ransome Alvin	1979-1984	Maintenance
GULLIXSON, Craig Alan	1983-1991	Instrument maker
HALL, John Scoville	1958-1977	Astronomer, Director
HAMILTON, George Hall	1917-1922	Planetary astronomer
HARDIE, Robert Howle	1953-1955	Astronomer
HARDWICK, Gloria Griffith	1988-1991	Public program
HARGRAVES, Donald	1931-1933	Astronomer (Visiting)
HART, Mary Ellen	1966-1968	Assistant
HESS, Seymour	1948-1950	Meteorologist (contract)
HILL, Beulah C.	1903-1905	Secretary
HOAG, Arthur Allen	1977-1986	Astronomer, Director
HOLTZMAN, Jon Andrew	1990-----	Astronomer
HORSTMAN, Helen Sylvia	1964-----	Secretary
HOYT, William Graves	1981-1985	Historian
HUNTER, Deidre Ann	1986-----	Astronomer
HUSSEY, Harry	1900-1905, 1915	Dome porter
INGE, Jay Landon	1966-1974	Illustrator
INGE, Mimi Mu Land	1985-----	Bookkeeper
IRIARTE, Braulio	1958-1960	Astronomer (visiting)
JENNINGS, H. J.	1928-1936	Maintenance
JERZYKIEWICZ, Mikolaj	1963-1966	Astronomer
JOHNSON, Harold Lester	1948-1959	Astronomer
JONES, Stuart Edgar	1962-1990	Photographer
JONES, "Skinny"	1894-1916	Carpenter
KANTZ, John Joseph	1978-1981	Instrument maker
KENT, Bernice Frances (Mrs. Giclas)	1931-1932	Secretary
KENT, Charles	1948-1950	Maintenance
KEPHART, Susan Beaver	1976-1979	Librarian
KNUCKLES, Claude	1954-1956	Observer (contract)

KRAMER, Katherine Xavier	1990-----	Fund raiser
KREIDL, Tobias Joachim	1980-1993	Astrono./computer programmer
KRZEMINSKI, Voytek	1961-1962	Astronomer
LABER, Patricia S.	1958-1961	Secretary
LAMPLAND, Carl Otto	1902-1951	Astronomer
LAUGHEAD, Charles E.	1967-1969	Programmer
LAWSON, Earl John	1964-1966	ACIC Observer
LEONARD, Edith H.	1894-1896	Secretary
LEONARD, Wrexie Louise	1896-1916	Secretary
LEWIS, Glenn Howard	1991-----	Volunteer
LLOYD, Richard Edwards	1962-1966	Computer assistant
LOCKWOOD, George Wesley	1973-----	Astronomer
LOPEZ, Arnold Esparza	1963-1968	Grounds care
LUMME, Kari A.	1979-1983	Astronomer (visiting)
LUTZ, Barry Lafean	1977-1991	Astronomer
MacDONALD, John Kenneth	1903-1916	Computer - Planet X
MAKINSON, J. W.	1908-1909	Research assistant
MARTIN, Leonard James	1967-----	Research assistant
MAULFAIR, Robert J.	1965-1968	Technician ACIC
McCANN, Terence Clifford	1963-1969	ACIC illustrator
McGLOTHLIN, Gerald	1975-----	Maintenance
MILLIS, Robert Lowell	1967-----	Astronomer, Director
MILLS, Edward C.	1915-1932	Carpenter
MITCHELL, Richard Irwin	1955-1960	Astronomer
MITCHELL, Thomas F.	1965-1968	ACIC illustrator
MUINONEN, Karri	1990-1992	Astronomer (visiting)
MULLINS, Kevin Francis	1976-1979	Maintenance
NEUMAN, Kenneth A.	1930-1933	Observer/assistant
NICHOLAS, Lilla D.	1965-1968	ACIC Secretary
NIERO, Lois	1972-1976	Secretary
NOTMAN, Frederick W.	1896-1916	Accountant (Boston)
NYE, Ralph Arthur	1976-----	Designer/engineer
O'CONNER, Johnson	1912-1914	Computer Planet X
OLIVER, Richard Charles	1980-----	Electronic technician
OSTERBERG, Charles	1948-1953	Photographic assistant
PANOFSKY, Hans Arnold	1948-1949	Meteorologist (contract)
PETERS, George	1931-1933	Visiting observer
PRATT, Susan Deanne	1966-1968	Darkroom assistant
PRISER, Michael John	1960-1964	Observer (contract)
PURGATHOFER, Alois	1961-1962	Astronomer
RAKOS, Karl D.	1963-1968	Astronomer (visiting)
RILEY, Louise Anderson	1965-1978	Observer assistant
ROETHER, Leroy L.	1963-1968	Groundskeeper
ROQUES, Paul Eugene	1987-1993	Public program/Photo lab
ROST, Kurt	1966-1969	Electrical engineer
SCHAAL, G. C.	1925-1932	Maintenance
SCHEELE, Helen Jean	1957-1985	Bookkeeper
SCHLEICHER, David Glenn	1985-----	Astronomer
SCHLOSS, George William	1972-1978	Craftsman
SCHMIDT, Stephen Daniel	1991-----	Public program
SCHONER, Steven R.	1971-1974	Public program
SEE, Thomas Jefferson Jackson	1896-1898	Astronomer
SEEGLITZ, Albert Henry	1958-1960	Assistant (contract)

SERKOWSKI, Krzysztof	1959-1966	Astronomer
SHANKS, Donald Demmon	1954-1985	Instrument maker
SHANKS, Janet Marie	1966-1968	Computer assistant
SINTON, William Merz	1957-1967	Astronomer
SINTON, Marjorie Korner	1959-1961	Research assistant
SKIFF, Brian Allan	1980-----	Research assistant
SLAUGHTER, Charles D.	1957-1959	Proper Motion Study assistant
SLIPHER, Earl Charles	1906-1964	Astronomer, Director
SLIPHER, Vesto Melvin	1901-1954	Astronomer, Director
SNYDER, Clifford E.	1965-1969	ACIC Illustrator
SOLLER, Theodore	1967-1975	Maintenance
SPENCER, John Robert	1991-----	Astronomer
STABLEFORD, Charles	1951-1954	Assistant
SYKES, Godfrey	1894-1910	Builder
SYKES, Stanley	1894-1955	Instrument maker
TERMANSEN, G. J.	1908-1909	Research assistant
THOMAS, Norman Gene	1958-1987	Proper Motion Study assistant
THOMSEN, Bjarne	1979-1983	Visiting astronomer
THOMPSON, Don Thomas	1969-----	Research assistant/editor
THOMPSON, Mina May	1987-1992	Secretary
TIFFT, William Grant	1961-1964	Astronomer
TOMBAUGH, Clyde William	1929-1945	Astronomer
TOMBLER, Marie	1954-1960	Secretary
TRUMAN, Orley Hosmer	1919-1924	Research assistant
VIELMA, Ramon	1931-1948	Maintenance
VIGIL, Barbara	1965-1968	Technician
VILIBORGHI, Mark R.	1970-1974	Electronic technician
WASSERMAN, Lawrence Harvey	1974-----	Astronomer
WATERBURY, George A.	1897-1898	Research assistant
WHITE, Nathaniel Miller	1969-----	Astronomer
WILLIAMS, Elizabeth Langdon	1909-1922	Chief computer - Planet "X"
WILLIAMS, Kenneth Leaman	1967-1975	Computer programmer
WILLIAMS, Kenneth Powers	1907-1908	Astronomer (Fellow)
WILSON, Albert George	1954-1957	Astronomer, Director

The Explorers of Mars Hill / 268

Chronology of Landmark Achievements

**1902
to 1914**   **Lowell, Williams & Macdonald**
Initiation of Planet "X" calculation by the study of unexplained perturbations in the orbit of Uranus, not accounted for by the attraction of Neptune.

1903   **Lowell**
Extensive studies of the waxing and waning of polar frost caps on Mars, of the layering of ice clouds in the Martian atmosphere, and of the global distribution of such cloudiness.

1909   **V. M. Slipher**
Correct interpretation of "stationary" spectral absorption lines of calcium and sodium as being due to the presence of gas clouds between stars.

1912   **V. M. Slipher**
First reliable measurements of the recessional velocities of galaxies; a key step in the development of the currently accepted idea of an expanding Universe.

1914   **V. M. Slipher**
Early measurements of the rotations of galaxies, suggesting they are huge systems with large masses.

1916   **V. M. Slipher**
Study and interpretation of bright emission lines from the Earth's upper atmosphere.

1921   **Lampland**
Discovery and extended observation of a giant gas shell around the dying star R Aquarii, an early hint at the key role played by mass loss from dying red stars.

1929   **V. M. Slipher & Tombaugh**
Reinitiation of the search for Planet "X" with the "Pluto Telescope," culminating a year later in the finding of the ninth planet in the solar system.

1936   **Adel**
Discovery of transparent "windows" in the atmosphere of the Earth, through which ground-based infrared astronomical measurements are now made.

1955   **Walker & Hardie**
Determination of the rotation period of Pluto.

1959   **Sinton**
Spectrographic hints of vegetation on Mars, later spurring biological scientists and NASA Viking missions, which found no evidence of life forms.

1960   **Sinton, Geoffrion & Korner**
Observation and mapping of lunar temperatures, indicating variations with time of lunar day and with the albedo of the lunar surface.

1960   **Sinton**
Temperature scans of lunar surface during eclipse, which showed younger craters (Tycho) warmer than the rest of the lunar surface - thinner dust layering.

1961   **Johnson, Hoag, Iriarte, Mitchell & K. Hallam**
Important studies of 106 galactic clusters, confirmed theories of stellar evolution, gave the distance to the clusters, indicated their age and a measure of the interstellar extinction (dust) between Earth and the cluster.

1962   **Sinton**
Discovery of CO in the atmosphere of a star (Betelgeuse) and in that of a planet, Venus.

1957
to 1978   Giclas
           "Lowell Proper Motion Survey." Discovery and cataloging of over 12,000 nearby
           and high velocity stars — identifying 1,800 new white dwarf suspects, 5 new
           comets and over 800 minor planets and compilation of a major astronomical
           sourcebook.

1973   Inge
           Extensive mapping of Martian albedo features, showing changes from year to
           year and how they relate to topography.

1974   Martin
           First definitive study of the extraordinary dust storms that periodically engulf
           Mars, showing when and where they start and how they move and spread.

1976   Baum & R. F. Nielsen
           Observational confirmation that the fundamental physical constants of nature are
           the same in distant parts of the Universe as they are on Earth.

1977   Millis, Wasserman, Franz et al
           Successful application of the occultation technique to determination of a number
           of properties of planets and other solar system bodies. Discovery of the rings of
           Uranus, measurement of Pluto's extended atmosphere and determination of the
           sizes and shapes of asteroids.

1979
to 1991   Bowell
           Determination of the orbits of large numbers of asteroids, enabling the charting
           of the dynamical structure of the asteroid belt. Identification of the first Mars
           Trojan asteroid.

1981   Franz & H. A. McAllister
           Collaboration in the application of speckle interferometry to overcome the blur-
           ring effects of the Earth's atmosphere. Improved knowledge on the incidence of
           double and multiple systems in stellar populations; improved accuracy of binary
           star observations and thus in determination of stellar masses.

1981   Baum & Thomsen
           First direct measurements of chemical abundance levels in the outer reaches of
           spheroidal galaxies offering data on the formation of such galaxies.

1981   Franz
           Reliable determination of the brightness and orbital brightness variation of
           Neptune's moon, Triton.

1984   White
           Definition and completion of a program of lunar occultations to measure sizes
           and atmospheric structures for a large sample of stars.

1984   Gallagher
           Demonstration that rapidly evolving, blue galaxies exist at the present time, com-
           plicating their use as "clocks" to measure the rate of evolution of galaxy popula-
           tions in the Universe.

1984   Baum & Kreidl
           Discovery of nature and extent of a tenuous ring around Saturn far outside its
           well-known bright rings.

1986   Lockwood & Thompson
           Confirmation of a direct connection between levels of sunspot activity and condi-
           tions in the atmosphere of Neptune, showing that solar activity can influence the
           atmospheres of planets, including Earth.

1986	**Millis & Schleicher**
	Discovery of periodic variability in the activity of Comet Halley, thus stimulating widespread research on the dynamics of comet nuclei and significantly affecting the interpretation of both ground-based and spacecraft data.
1987	**Bus**
	Discovery of periodic variability in the brightness of the enigmatic object (2060) Chiron.
1987 to 1992	**Hunter**
	Discovery that the upper stellar mass limit is not correlated with the mass of the host star-forming molecular cloud. Discovery of very large filaments of ionized gas in irregular galaxies far from any apparent stellar sources of ionizing radiation.
1988	**Lockwood & Skiff**
	Measurements of micro-brightness fluctuations in stars similar to the Sun which are due to "sunspots" on these stars leading to determinations of reliability and "normality" of the Sun.
1988	**Lutz & T. C. Owen**
	Collaboration in the measurement of deuterium oxide on Mars, implying that Mars is likely to have lost a large amount of ordinary water from its atmosphere.
1989	**Kreidl**
	Discovery of peculiar stars that appear to vibrate like ringing bells - offering prime targets for probes of their internal structure via stellar seismology.
1990	**Bus et al**
	Discovery of CN emission band in the spectrum of the "asteroid" (2060) Chiron, confirming that object's actual cometary nature.
1992	**Holtzman et al**
	Discovery with Hubble Space Telescope of blue, presumably young globular clusters in NGC 1275.

271 / *Appendices*

From Percival to Percival

John Lowell (1595-1647), son of Percival Lowle, had a wife and children of his own when he accompanied his father to the New World in 1639. He was elected constable of Newbury in 1641, town clerk the following year, a member of the Massachusetts General Court (colonial legislature) in 1644 and magistrate in 1645. In 1629, his first wife, Mary, had presented him with a son, also named John, a name that has stayed singular and redundantly prominent within the Lowell family even to the present day.

The second John became a cooper (barrel-maker) by trade, migrated to the less fashionable "South Shore"[1] and finally died in the colonial capital of Boston in 1694. His son, Ebenezer (1675-1711), the oldest child of John's third marriage (to Naomi Sylvester of Scituate) became a prosperous cordwainer, a long lost occupation which dealt with the importing and processing of leather, and soon found himself heavily involved with shipping from the port of Boston. His marriage to Elizabeth Shailer, in the year of his father's death, produced several children, of which the first American-born John saw light in March 1704.

Ordained to the Congregational clergy in 1726, after training at Harvard College, John returned to his grandfather's home town of Newbury to become pastor of its newly established Third Church, where he remained as it spiritual leader until his death forty-one years later.

A fourth John was born in 1743 of the Reverend's 1725 marriage to Sarah Champney. Known in subsequent family records as "the old judge" (to distinguish him from yet another John, his great grandson, who also served in the Federal judiciary), he became the cornerstone of the family's subsequent prominence in both New England society and in its business community, the two being largely intertwined. A man of great mathematical ability, the old judge became a founder of the American Academy of Arts and Sciences, a delegate to the Continental Congress, an appointee by George Washington to the bench and a perennial member of the Corporation of Harvard College. The old judge (1743-1802) was a man of many interests and great prestige; from his day forward, his descendants were to exhibit considerable mathematical skill and were also almost unanimously well travelled, well read and well off.

John's first marriage was to Sarah Higginson in 1767, from which union came another John (1769-1840), later to be known by the political writings of his own pen as "the Rebel." Sarah died in 1772 while her husband was still residing in Newburyport and, though not yet a citizen of prominence, John was a virile man with young children to bring up. Two years later he was married again, to Susanna Cabot of Salem. From this second union came a second branch of the Lowell family, famous for such stars as Francis Cabot Lowell (1775-1817), America's pioneer industrialist and the major founder of the family's ongoing prosperity.

When Susanna died in 1777, the future judge married for a third time, to Rebecca Russell Tyng, the thirty-year-old widow of James Tyng of Dunstable and daughter of Massachusetts judge James Russell (1715-1798). Of this union came three children with which this chronology deals in part with Charles Lowell (1782-1861) who followed tradition by serving as pastor of Boston's West Church from 1806 until his death. While Percival, the astronomer, was descended from the senior line of the old judge's progeny, his younger sister, Elizabeth, was to marry the eldest child of Harriet, daughter of Charles Russell Lowell (1807-1870), and a great-great-grandson of the old judge by his third wife. It was to this third cousin (and brother-in-law) that Percival turned for help in his times of need

---

1. In Bay State parlance, there are only two shores relative to the ocean - one of them can be found to the north of Boston and the other lies to the south. That to the south was largely populated by descendants of those from the "Old Colony" - i.e. Plymouth, and thus not nearly of the same social status as many of the more gentrified emigrants who settled in Boston or areas to the north.

and who drew up for him the unique, bitterly contested will under the terms of which his observatory continues to function.

The Rebel, sickly as a child but endowed with great perception, married Rebecca Amory in 1793, from which union came another shrewd judge of business opportunity as well as custodian of the public interest, John Amory Lowell (1781-1881). It was he who managed the legacy under a will (drawn up on the banks of the Nile river) by John Lowell, Jr. (1799-1836), son of Francis Cabot Lowell. As the first sole trustee of his cousin's estate, John Amory Lowell established New England's unique Lowell Institute on its path to greatness and also established the concept of sole trusteeship in the family thought process. The successor trustees of the Lowell Institute have been his son, Augustus; his grandson, Lawrence; his great-grandson, Ralph; and his great-great-grandson, John.

John Amory Lowell's first marriage was to his cousin, Susan Cabot Lowell and produced a further line, of which the eldest was John (1824-1897), appointed to the Federal bench by Abraham Lincoln. This line also included John's architect grandson, Guy, who carried the burden of preserving Percival's estate from the grasping hands of his widow. John Amory Lowell's second marriage, in 1829, was to Elizabeth Cabot Putnam, daughter of Judge Samuel Putnam (1768-1853), who remains famous in legal circles for his "prudent man" doctrine on the administration of estates. The eldest child of this union was Augustus (1830-1901), who in turn became the custodian in their final years of the family's business interests in the city named for his great uncle, Francis Cabot Lowell.

273 / *Appendices*

Selective Astronomical Glossary

ALBEDO—The fraction of incident sunlight that a planet reflects.

ANGULAR DIAMETER—Angle subtended by the diameter of an object.

APHELION—The point in its orbit where a planet is farthest from the sun.

APPARENT MAGNITUDE—A measure of the observed light received at the Earth from a star or other object.

ASTRONOMICAL UNIT (AU)—Meant to be the semi-major axis of the orbit of the Earth; now defined as "the semi-major axis if the orbit of a hypothetical body with the mass and period that Karl Gauss assumed for the Earth"; the semi-major axis of the Earth's orbit is 1.000,000,230 AU's.

ATMOSPHERIC REFRACTION—The bending of light rays from celestial objects by the Earth's atmosphere.

AURORA—Light radiated by the ionosphere, mostly in the polar regions.

AZIMUTH—The angle along the observer's horizon, measured eastward from the north, to the intersection of the horizon with the vertical circle passing through an object—i. e. a numerical value of the object's direction.

BANDS (in spectra)—A group of emission or absorption lines, usually so numerous and closely spaced that they coalesce.

BINARY STAR—A double star; two stars revolving about each other.

BLINK MICROSCOPE—A microscope in which the user's view is shifted rapidly back and forth between the corresponding portions of two different photographs of the same region of the sky.

BODE'S LAW—A sequence of numbers that give the approximate distances of the planets from the Sun.

BOLIDE—A very bright fireball or meteor.

CELESTIAL EQUATOR—The circle of intersection of the celestial sphere with the plane of the Earth's equator.

CELESTIAL MECHANICS—That branch of astronomy dealing with the motions and gravitational influences of solar, stellar and planetary bodies.

CENTER of MASS—The mean position of the various mass elements of a body or sytst weighted according to their distance from that center of mass.

CEPHEID VARIABLE—a form of yellow supergiant pulsating star.

CERES—the largest of the asteroids and the first to be discovered.

CHROMATIC ABERRATION—A defect of optical systems whereby light of different colors is focussed at different places.

CONJUNCTION—The condition when a planet has the same celestial longitude as the Sun; a condition when two celestial bodies have the same longitude or right ascension.

CONSTELLATION—A grouping of stars named for a particular object; the area of the sky assigned to such a configuration.

CORONA—The outer atmosphere of the Sun.

COSMIC RAY—Atomic nucleus that strikes Earth's atmosphere with high energy.

COSMOGONY—The study of the origin of stellar or cosmic bodies.

DECLINATION—Angular distance of some object north or south of the celestial equator.

DIURNAL CIRCLE—The apparent path of a star in the sky during a complete terrestrial day.

DOPPLER SHIFT—The apparent change in wavelength of the radiation from a source due to its relative motion towards or away from the place of observation.

DYNE—The metric unit of force required to accelerate a mass of one gram one centimeter per second per second.

ECLIPTIC—The apparent annual path of the sun on the celestial sphere.

ECCENTRIC—The point, about which an object revolves, that is not at the center of the circle.

ECCENTRICITY (of ellipse)—The ratio of the distance between the foci to the major axis.

ECLIPSE—The cutting off of all or part of the light of one body by another passing in front of it.

ECLIPSE PATH—The track along the Earth's surface swept by the shadow of the moon.

ELLIPSE—A conic section; the curve of intersection of a circular cone and a plane cutting completely through the cone.

ELLIPTICITY—The ratio (in an ellipse) of the major axis to the minor axis.

EOLIAN—From Aeolus, Greek god of the winds—wind related.

EPHEMERIS—A table that gives the positions of celestial bodies at various times, and other astronomical data.

EQUATORIAL MOUNT—A mounting for a telescope, one axis of which is parallel to the Earth's axis, so that a motion of the telescope about that axis can compensate for the Earth's rotation.

EQUINOX—One of the two intersections of the ecliptic and celestial equator.

FLARE—A sudden and temporary outburst of light from an extended region of the solar surface. Flare stars are those which alter dramatically in brightness.

FOCAL LENGTH—The distance from a lens or mirror to the point where the light it converges comes to a focus.

FOCAL RATIO (SPEED)—Ratio of the diameter or aperture of a lens or mirror to its focal length.

FOCUS—Point where the rays of light converged by a mirror or lens meet.

GALACTIC CLUSTER—A cluster of stars located in the spiral arms or disk of a galaxy.

GALAXY—A large assembly of stars; a typical galaxy contains hundreds of billions of stars.

GAMMA RAYS—Photons (of electromagnetic radiation) of energy higher than x-rays; the most energetic form of electro-magnetic radiation.

GRAVITATION—The tendency of matter to attract itself.

GREAT CIRCLE—Circle on the surface of a sphere that is the curve of intersection of the sphere with a plane passing through its center.

GREENWICH MERIDIAN—The meridian of longitude passing through the site of Greenwich Observatory, near London.

HYDROSTATIC EQUILIBRIUM—A balance between the weights of various layers, as in a star or the Earth, and the pressures that support them.

INDEX OF REFRACTION—A measure of the refracting power of a transparent substance; the ratio of the speed of light in a vacuum to its speed in the substance.

INSOLATION—The rate at which radiation from the Sun is received per unit area on the ground.

INTERFEROMETER—A device, making use of the principle of interference of waves, with which small angles can be measured.

ION—An atom that has become electrically charged by the addition or loss of one or more electrons.

IONIZATION—The process by which an atom gains or loses electrons.

IONOSPHERE—The upper region of the earth's atmosphere in which many of the atoms are ionized.

KEPLER'S LAWS—Three laws, discovered by Johan Kepler, that describe the motions of the planets about the Sun.

KINETIC ENERGY—Energy associated with motion; the kinetic energy of a body is one half the product of its mass and the square of its velocity.

LIGHT YEAR—The distance light travels in a vacuum in one year; approximately 6,000,000,000,000 miles.

LIMITING MAGNITUDE—The faintest magnitude that can be observed with a given instrument under given conditions.

LINEAR DIAMETER—Actual diameter in units of length.

LUMINOSITY—The rate of radiation of electromagnetic energy into space by a star or other object.

MARE—Latin for "sea"; a name applied to many of the "sea-like" features observed on the Moon or Mars.

MASS—A measure of the total amount of material in a body, defined either by its inertial properties or gravitational influence.

MERIDIAN (CELESTIAL)—The great circle on the celestial sphere that passes through an observer's zenith and the celestial pole.

METEOR—The luminous phenomenon observed when a meteoroid enters the Earth's atmosphere and burns up - popularly called a "shooting star."

MINOR AXIS—The smaller diameter of an ellipse.

NADIR—The point on the celestial sphere directly opposite the zenith.

NAUTICAL MILE—The mean length of one minute of arc on the Earth's surface along the equator.

NEBULA—A cloud of interstellar gas or dust.

NODE—The intersection of the orbit of a body with a fundamental plane - usually that of the celestial equator or of the ecliptic.

OBJECTIVE—The principal image-forming component of a telescope or other optical instrument.

OBLATE SPHEROID—A solid formed by rotating an ellipse about its minor axis; planet Earth is one.

OBLATENESS—A measure of the flattening of a sphere due to centrifugal force.

OBLIQUITY OF THE ECLIPTIC—The angle between the planes of the celestial equator and the ecliptic; presently about $23\,{}^{1}\!/_{2}°$.

OCCULTATION—An eclipse of a star or planet by the Moon or another planet.

ORBIT—The path of a body that is in revolution about another body.

PARABOLA—The curve of intersection between a circular cone and a plane parallel to its axis.

PARALLAX—An apparent displacement of an object due to motion of the observer.

PARSEC—The distance of an object that would have a stellar parallax of one second of arc when the observer moves the diameter of the earth's orbit; 1 parsec = 3.26 light years.

PENUMBRA—The portion of a shadow from which only a portion of the light source is occulted by an opaque body.

PERTURBATION—The disturbing effect, when small, produced by an external agent on the motion of a body as predicted by simple two-body theory.

PERIHELION—The place in the orbit of an object revolving about the Sun where it is closest to the center of the Sun.

PERIOD—The time required for one complete revolution of a rotating body about its parent; the time required for a recurring event to repeat.

PHOTOMETRY—The measurement of light intensities.

PHOTOMULTIPLIER—A photoelectric cell in which the electric current generated by incoming light is amplified.

PHOTON—A discrete unit of electromagnetic energy.

POLAR AXIS—The axis of rotation of the Earth; an axis in the mounting of a telescope that is parallel to the axis of the Earth.

PRIME FOCUS—The point in a telescope where the objective focusses the light.

PRIME MERIDIAN—The terrestrial meridian passing through the site of the Greenwich Observatory - Longitude 0°.

PROPER MOTION—The angular movement of a star per year.

RADIATION—A mode of energy transport through a vacuum; such energy itself.

RED SHIFT—A shift to longer wavelength of the light from remote galaxies, presumed to be a Doppler shift caused by their motion away from Earth.

REFLECTING TELESCOPE—A telescope in which the principal optical component is a concave mirror.

REFRACTING TELESCOPE—A telescope in which the principal optical component is a lens or system of lenses.

REFRACTION—The bending of light rays passing through a transparent medium.

RESOLVING POWER—A measure of the ability of an optical system to resolve or separate fine details in the image it produces.

RETROGRADE MOTION—An apparent westward motion of a planet on the celestial sphere or with respect to the stars.

RIGHT ASCENSION—A coordinate for measuring the east-west position of celestial bodies.

SCHMIDT TELESCOPE—A type of reflecting telescope in which certain aberrations produced by the objective mirror are compensated for by a thin corrective lens.

SOLAR CONSTANT—The mean amount of solar radiation received, at the distance of the Earth, per unit time normal to the direction of the Sun.

SOLAR WIND—A radial flow of corpuscular radiation leaving the Sun.

SOLSTICE—The points on the celestial sphere where the Sun reaches its maximum distances north and south of the celestial equator; the time when the Sun reaches these points.

SPECTROGRAPH—An instrument for photographing a spectrum; usually attached to a telescope.

SPECTROHELIOGRAPH—An instrument for photographing the sun, or part of the sun, in the monochromatic light of a particular spectral line.

SPECTROPHOTOMETRY—The measurement of the intensity of light from a star or other source at different wavelengths.

SPECTROSCOPE—An instrument for directly viewing the spectrum of a light source.

SPECTRUM—The array of colors or wavelengths obtained when light from a source is dispersed, as in passing through a prism or grating.

SPHERICAL ABERRATION—A defect of optical systems whereby on-axis rays of light striking different parts of the objective do not focus at the same place.

SPIRAL GALAXY—A flattened, rotating galaxy with pin-wheel like arms of interstellar material and young stars winding out from its nucleus.

277 / Appendices

SUNSPOT CYCLE—The semi-regular eleven-year period with which the frequency of sunspots fluctuates.

TELLURIC—Of terrestrial origin.

TELLURIC LINES—Lines or bands in the Sun's spectrum produced by absorption of sunlight in the Earth's atmosphere.

TERMINATOR—The line of sunrise or sunset on the moon or planet.

THERMOCOUPLE—A device for measuring the intensity of infrared radiation or radiant heat.

TRANSIT—An instrument for timing the exact instant a star or other object crosses the local meridian. Also, the passage of a celestial body across the meridian; or the passage of a small body across the disk of a larger one.

UMBRA—The completely dark part of a shadow.

ZENITH—The point on the celestial sphere opposite the direction of gravity; the point directly overhead.

ZODIAC—A belt around the sky centered on the ecliptic.

ZODIACAL LIGHT—A faint illumination along the zodiac, believed to be sunlight reflected and scattered by interplanetary dust.

Selective Planetary Data

PLANET	SEMI-MAJOR AXIS (IN AUs)	SIDEREAL PERIOD (IN YEARS)	DIAMETER IN KM	ECCENTRICITY	INCLINATION TO ECLIPTIC
MERCURY	0.3871	0.24085 (87.97 DAYS)	4,680	0.20563	7.004°
VENUS	0.7233	0.61521 (224.7 DAYS)	12,200	0.00679	3.394
EARTH	1.0000	1.000039 (365.26 DAYS)	12,742	0.01673	
MARS	1.5237	1.88089 (686.98 DAYS)	6,648	0.09337	1.850
JUPITER	5.2037	11.8653	139,500	0.04863	1.306
SATURN	9.5803	29.6501	116,340	0.05099	2.487
URANUS	19.141	83.7445	47,500	0.04579	0.772
NEPTUNE	30.1982	165.951	44,800	0.00456	1.773
PLUTO	39.4387	247.687	6,000	0.25024	17.17

Bibliography

Published books pertinent to Percival Lowell
and those connected with his Observatory

Damon, S. Foster, *Amy Lowell, A Chronicle*. Houghton, Mifflin, Boston, 1935.

Greenslet, Ferris, *The Lowells and Their Seven Worlds*. Houghton, Mifflin, Boston, 1946.

Hoyt, William Graves, *Lowell and Mars*. University of Arizona Press, Tucson, 1976.

Hoyt, William Graves, *Vesto Melvin Slipher (1875-1969), Biographical Memoirs* (Vol. 52). National Academy Press, Washington, DC, 1980.

Hoyt, William Graves, *Coon Mountain Controversies*. University of Arizona Press, Tucson, 1987.

Hoyt, William Graves, *Planets X and Pluto*. University of Arizona Press, Tucson, 1980.

Leonard, Wrexie Louise, *Percival Lowell - An Afterglow*. Richard Badger, Boston, 1921.

Levy, David H., *Clyde Tombaugh - Discoverer of Pluto*. University of Arizona Press, Tucson, 1991.

Lowell, A. Lawrence, *Biography of Percival Lowell*. MacMillan, New York, 1935.

Lowell, Delmar, *The Historical Genealogy of the Lowells of America From 1639 - 1891*. Rutland, VT, 1899.

Lowell, Percival, *Choson, The Land of Morning Calm - A Sketch of Korea*. Ticknor & Co., Boston, 1886. (2 reprints)

Lowell, Percival, *The Soul of the Far East*. Houghton, Mifflin, Boston, 1888. (6 reprint editions - 1 German edition)

Lowell, Percival, *Noto - An Unexplored Corner of Japan*. Houghton, Mifflin, Boston, 1891. (reprint 1895 - Japanese edition 1979)

Lowell, Percival, *Occult Japan - Or the Way of the Gods*. Houghton, Mifflin, Boston, 1894. (reprint 1895)

Lowell, Percival, *Mars*. Houghton, Mifflin, Boston, 1895. (3 reprint editions - 1 British - 1 Chinese)

Lowell, Percival, *The Solar System*. Houghton, Mifflin, Boston, 1903. (reprint, 1908)

Lowell, Percival, *Mars and its Canals*. MacMillan, New York, 1906. (3 reprint editions - 1 German - 1 French - 1 Swedish)

Lowell, Percival, *Mars As the Abode of Life*. MacMillan, New York, 1908. (1 reprint)

Lowell, Percival, *The Evolution of Worlds*. MacMillan, New York, 1909. (1 reprint)

Morse, Edward Sylvester, *Mars and Its Mystery*. Little, Brown, Boston, 1906.

Putnam, W. L., *A Yankee Image - The Life and Times of Roger Lowell Putnam*. Phoenix Publishing, West Kennebunk, ME, 1991.

Sheehan, William, *Planets and Perceptions: Telescopic Views and Interpretations—1609 - 1909*. University of Arizona Press, Tucson, 1988.

Sykes, Godfrey, *FRGS, A Westerly Trend*. Arizona Historical Society & University of Arizona Press, Tucson, 1945. "A Veracious Account of His Travels."

Webb, George Ernest, *Tree Rings and Telescopes*. University of Arizona Press, Tucson, 1983. The scientific career of A. E. Douglass.

Weeks, Edward, *The Lowells and Their Institute*. Little, Brown, Boston, 1966.

Yeomans, Henry Aaron, *Abbott Lawrence Lowell - 1856-1943*. Harvard University Press, Cambridge, 1948.

Articles

Statistical summary of published articles by Percival Lowell*

Korean and Japanese history and social customs	27
Other history	5
Current social issues & miscellaneous	19
Astronomy in general, including Solar System	66
Mars and its features	107
Venus and its features	10
Saturn and its features	16
Mercury and its features	7
Jupiter and its features	3
Uranus and its features	2

Of the above total, 22 were written in French and 12 in German or Swedish.

*Adapted from a compilation of scientific literature by David Strauss.

Illustrations and Credits

Although the great majority of the illustration in *Explorers of Mars Hill* came from the archives of Lowell Observatory, the author is deeply grateful to all those who supplied needed material from their own collections. In the following the abbreviated caption is followed by the source, the photographer's name where known or relevant, and the page number on which the illustration appears. In this listing the Observatory archives are abbreviated LOA. The front and back endleaf photos of Mars and Saturn respectively are used by courtesy of NASA.

Percival Lowell / LOA, x

Acrylic painting by East Flagstaff Junior High / LOA, xx

Clark telescope with Brashear spectrograph, 1903 / LOA, 2

Abbott Lawrence and Percival Lowell, 1859 / LOA, 4

Percival Lowell and "Tejiro," 1883 / LOA, 8

Percival Lowell in his Tokyo garden with Ralph Curtis, 1883 / LOA, 9

First Korean Trade Mission to the United States, 1884 / LOA, 10

Giovanni Virginio Schiaparelli, circa 1870 / LOA, 12

View of "Site Eleven" from Aspen Street, 1893 / LOA, 18

Pickering's 18-inch prefabricated telescope dome being assembled, 1894 / LOA, 19

Two borrowed telescopes on their single mounting, 1894 / LOA, 20

Bank Hotel in Flagstaff, 1892 / Courtesy Monte Vista Hotel, 26

Sykes 24-inch telescope dome in Tacubaya, Mexico, 1896 / LOA, 27

Mexican dome porter winding clock weights at Tacubaya / LOA, 28

Mars Hill with the Clark dome in place, 1897 / LOA, 29

Douglass sketching at the 24-inch dome, 1898 / LOA, 30

Cogshall and See at 24-inch telescope, 1897 / LOA, 32

The Clark dome complete, 1900 / LOA, 35

Pontoons being assembled for the Clark dome, 1899 / LOA, 36

Percival Lowell on the lecture circuit, 1896 / LOA, 39

William Lowell Putnam and his wife, 1902 / LOA, 40

Slipher family home, 1920 / LOA, 45

John Miller and Albert Einstein, 1929 / LOA, 46

Observatory staff at the 24-inch dome, 1905 / LOA, 50

Carl Lundin cleaning the lens for the 24-inch telescope, 1901 / LOA, 51

Percival Lowell and Venus, his cow, 1905 / LOA, 52

V.M. Slipher studying Triassic dinosaur footprint, 1928 / Marcia Slipher Nicholson collection / LOA, 58

Building the Monte Vista Hotel, 1926 / Courtesy Monte Vista Hotel, 59

Marcia Slipher's wedding party, 1931 / Allan Cree collection / LOA, 60

V.M. and E.C. Slipher, 1936 / LOA, 61

V.M. Slipher in discussion with Lampland, 1947 / LOA, 62

Arthur Adel at the 24-inch telescope, 1936 / LOA, 63

Retirement party for V.M. Slipher, 1954 / LOA, 64

Henry Giclas and Albert Wilson with gifts / LOA, 65

Dr. and Mrs. Percival Lowell's wedding portrait, 1908 / LOA, 68

Wrexie Louise Leonard, 1895 / Mark Klauk collection / LOA, 70

Wrexie Leonard on Mars Hill, 1905 / LOA, 71

Observatory staff and servants at Baronial Mansion, 1905 / LOA, 72

Wrexie Leonard's bedroom, 1906 / LOA, 73

Pumpkin from Lowell's garden, 1916 / LOA, 74

The Explorers of Mars Hill / 282

283 / Appendices

Index

285

287 / *Index*

National Aeronautics & Space Ad. 190, 222
  Science Foundation xviii, 129, 171, 190
Neptune, planet 173
Newbury, town 3
Newcomb, S. xiv
Nicholson, S. B. 186
Nielsen, R. F. 270
Noble, G. W. C. 8
Notman, F. W. 74
Noto, peninsula 11
Nutter, G. R. 102

Oak Creek 122, 246
O'Conner, J. 174
Occultation studies 231
Office of Contract Settlement 166
Ohio State University 170
Ohio Wesleyan University 170
Ontake, mountain 11
Osgood, H. L. 75
Osterbrock, D. E. 47

Package Machinery Company 160
Padre Butte 169
Panofsky, H. A. 196
Parks, Rev. L. 82
Perkins, Hon. F. W. 95
  H. M. 129
  telescope 129, 167, 216
Perth Observatory 227, 228, 136
Pfund, A. H. 165
Phoenix, city 17
Photometry 191
Photomultipliers 192
Pickering, C. 13
  E. C. 13
  W. H. 13, 19, 23, 173, 187
Pierce, B. O. 182
Pioneer, spacecraft 230
Planet X (see Pluto)
Planetary Patrol 208, 228
  Research Center 152, 155, 158, 208, 226
Pluto, planet 60, 239
  discussion of 182
  naming of 161
  search for 158, 172
  telescope 120, 149, 151, 159, 176
Pollock, Hon. T. E. 126
Potter, E. C. 74
Prescott, city 17

Purgatoire River 113
Putnam, C. 6
  Col. D. 160
  Eliz. (see also Lowell, E.) 40, 157
  Geo. 6, 29, 87, 91
  H. 6
  Gen. I. 160
  Jos. 160
  M. C. J. 7, 146, 158, 218
  Hon. R. L. xviii, 7, 61, 119, 122, 157,
    176, 194, 217, 218, 255
  ancestry of 159
  Hon. S. 159
  W. L. II 6, 29, 36, 40, 41, 86, 99
  W. L. III 7, 146

Radial velocities 43
Randall, H. M. 164
Reaves, G. xi, 172
Red shift 44, 54
Reeve, J. S. 5
Reid, M. 112
Riordan, D. M. 18
Ritchey, G. W. 136
Roden, W. 114
Roques, P xi
Roosevelt, A. 5
  Hon. F. D. 161
  Hon T. 29
Ross, F. E. 167
Rotch, A. L. 38
Royal Astronomical Society 12, 60, 164
  Photographical Society 201
Rubin, V. 169
Russell, H. N. 122, 253
  W. 6

Saa, O. 200
St. Bartholomew's Church 82
San Francisco Peaks 148
Sargent, C. S. 200, 246
  H. 92
  J. S. 94
Saturn, observations of 199, 229
Schall, G. 252
Schiaparelli, G. V. 12, 22, 172
Schottland, M. 171
Schultz Peak 120, 148
Seamans, A. L. xi
See, T. J. J. 26, 29, 32, 34
Seeing conditions 14
Serkowski, K. 148
Sevenels, house 7

The Explorers of Mars Hill / 288

The
Explorers
of Mars Hill

has been published in a first edition
of five thousand copies.
Designed by A. L. Morris,
the text was composed in Palatino
and printed by Knowlton & McLeary
in Farmington, Maine, on Champion Pageantry Text.
The jacket and endleaves were printed on
Curtis Tweedweave Text,
and the binding in Holliston Mills Roxite,
was executed by New Hampshire Bindery
in Concord, New Hampshire.